HOWARD HUGHES
AND THE SPRUCE GOOSE

Graham M Simons

First published in Great Britain in 2014 by
Pen and Sword Aviation

An imprint of
Pen & Sword Books Ltd
47 Church Street
Barnsley
South Yorkshire
S70 2AS

Copyright © Graham M. Simons, 2014

ISBN 9 781 7838315 55

A CIP catalogue record for this book is
available from the British Library

Printed and bound in England
By CPI Group (UK) Ltd, Croydon, CR0 4YY

Pen & Sword Books Ltd incorporates the Imprints of Pen & Sword Aviation, Pen &
Sword Family History, Pen & Sword Maritime, Pen & Sword Military, Pen & Sword
Discovery, Wharncliffe Local History, Wharncliffe True Crime, Wharncliffe
Transport, Pen & Sword Select, Pen & Sword Military Classics, Leo Cooper, The
Praetorian Press, Remember When, Seaforth Publishing and Frontline Publishing

For a complete list of Pen & Sword titles please contact
PEN & SWORD BOOKS LIMITED
47 Church Street, Barnsley, South Yorkshire, S70 2AS, England
E-mail: enquiries@pen-and-sword.co.uk
Website: www.pen-and-sword.co.uk

CONTENTS

ACKNOWLEDGEMENTS

A project of this nature could not be undertaken without considerable help from many organisations and individuals.

Special thanks must go to Col. Richard L Upstromm and Tom Brewer from the USAF Museum, now the National Museum of the USAF for the provision of photographs and details.

Background to the development and procurement of the project can be found in published and unpublished primary source research material in the form of memoranda, policy statements and other documents from the Army Air Corps, the USAF as provided by Lynn Gamma and all in the U.S. Air Force Historical Research Center at Maxwell Air Force Base, Montgomery, Ala. The same applies to the valuable services provided by the History Office of the Air Technical Service Command, Wright Field, Dayton, Ohio and the Public Affairs Office, Edwards Air Force Base. Much other primary source documentation is also located in the National Archives and Records Administration at College Park, Maryland.

The archives of the National Advisory Committee for Aeronautics provided access to all their relevant material, as did the archives of the Institute of Aircraft Production.

Special thanks should go to Simon Peters, for the use of his collection, which includes fragments of the archives of G Geoffrey Smith MBE, the managing editor of *Aircraft Production*.

Thanks should also go to Brian Cox, Martin Bowman and Peter Green for allowing me to dip into their respective photographic collections and archives. Personal thanks must also go to David Lee, the former Deputy Director and Curator of Aircraft of the Imperial War Museum at Duxford, to John Hamlin and to Vince Hemmings, the former and premier curator of the East Anglian Aviation Society's Tower Museum at Bassingbourn.

The author is indebted to many people and organisations for providing photographs for this book, many of which are in the public domain. However, despite strenuous efforts, in some cases it has not been possible to identify the original photographer and so credits are given in the appropriate places to the immediate supplier. If any of the pictures have not been correctly credited, the author apologises.

Introduction

'The Spruce Goose' was an epithet that not only Howard Hughes loathed, but was also factually wrong. Whether one calls it the HK-1, the H-4 or 'Hercules' - and it's possible to choose any one you wish, for they all are correct - but the general public will usually give you a blank look. Call it the Spruce Goose, and almost everyone knows what you are talking about - hence the title of this book!

And as for the spruce aspect, well, nothing could be further from the truth - it was however, a catchy nickname that had a ring to it.

Allegedly, it came about though the considerable controversy which was to surround its funding. After US Senator Ralph Owen Brewster dubbed the HK-1 a 'flying lumberyard,' the 'Spruce Goose' nickname was coined by the popular press of the day - and it stuck!

To fully comprehend the major role played by Howard Hughes in the story of the Spruce Goose, it is necessary to try to know this man as he really was - something that even when he was alive was very hard to do. Today, with the myriad of fictionalised and erroneous stories that have been written about him it is impossible to discover what is 'real' and what is Hollywood myth, and can be repudiated by facts. I do not intend to produce a work that is an attempt to picture a sainted Howard Hughes or a cover-up of his faults as a person; but rather to produce an honest effort to present the reader with the truth about one of the most remarkable men of the 20th century.

To accomplish this goal, consider this man's life and accomplishments unembellished by opinion and assumption. Then you will be able to clearly understand his personal growth and demise as the story of the Spruce Goose unfolds.

Howard Robard Hughes became known worldwide and yet was still a man of mystery to us because he was veiled in secrecy and seclusion. Many of those who knew him felt that they knew the 'real' Howard Hughes, but actually they knew only part of this extremely complex man - the part he chose to reveal. Knowing him in one facet of his life could never reveal the total man and I do not think that one person was ever privileged to know all aspects of his life.

Howard Hughes was clearly not the easiest of men to work for. Unlike so many employers he was an enthusiast - and what may have been much worse, he was a well-informed, very wealthy enthusiast. Aviation was from a very tender age by far and away his greatest love and interest, anyone who sold him an aeroplane had to meet a detailed specification which back in the 1930s was unheard of; and that was merely to provide Hughes with a basis for modification.

Hughes it seems had very little contact with any but a handful of people in his organisation. Yet in those contacts he was always pleasant to the people who worked for him on the lower levels - those who got their hands dirty. Working higher up in Hughes's organisation was worse; he would call in the middle of the night and talk for hours. One legend was that he regularly told his executives *'Look, the bankers and others I have to call during the day, but you work for me. I can call you any time.'*

It was often said that the executives lives were not their own. Sometimes he'd give them hell, and right in front of anyone else present. He'd ask a question and if they blustered and came out with the BS he'd tell them, *'If you don't know what you're talking about keep your goddamned mouth shut!'* Like most true engineers, all they had to say was 'I don't know,' and he would accept and respect that. but to so many executives, admitting such a thing was a complete anathama.

By all accounts Howard Hughes Jr was no typical industrialist, aviator, or anything else of the period - he was one of a kind. Hughes developed a passion for privacy, but not for anonymity. He wanted to be left alone, but at the same time he wanted to be widely known, respected, and recognised – but all on his terms.

He had a peculiar attitude of proprietorship over words and ideas and operated in secret and hidden ways. He formed the Rosemont Corporation to keep his name out of print, giving that organisation the sole right to any biographical material. Would-be biographers were either bought off or sued. His legal team drew up confidentially contracts that insisted on 'no talk' clauses for his employees. His Romaine Street command post in Hollywood was protected by twenty-four-hour guards, electronic gadgetry and warning devices.

Yet when it came to his pride and joy - the massive HK-1/H-4 'Hercules' flying boat - he could be lavish. 'Free to Use' photo-packs were distrubuted by Johnny Meyer, his press agent, to aviation orientated editors and every facility was provided to newsmen and journalists covering the activities.

Many people make much of his passion for secrecy that first bordered on and then dived deeply into paranoia. However, as we shall see, jealousy and personality clashes with high ranking members of the Army Air Corps, along with the nefarious activities and machinations of Senator Ralph Own Brewster and Pan American's Juan Terry Trippe plus the long-term health issues and painkiller addiction brought about by his crash in the XF-11 were enough to send anyone over the edge!

This then is the story of Hughes' massive flying boat - a machine conceived to meet a requirement that subsequently vanished; built and flown once by the determination of someone to prove that it could and preserved for thirty three years by that same person 'just because he could'.

It's been called a 'white elephant' and maybe it was, but the colour was silver not white, and no elephant looks as good as the Hercules!

Graham M Simons
Peterborough
July 2014

There Is A Need...

It was 1942 and America had been at war since 'the day of infamy' on 7th December 1942 when the Japanese attacked Pearl Harbor. The country had not yet recovered from the shock. It was a time of despair with continuing military setbacks in the Pacific, China, Africa and Russia. Even worse, the German U-boat (Unterseeboot, which means 'undersea boat') menace was claiming an ever-increasing toll of American and Allied shipping.

German U-boats had shown how dangerous they were from the very start of the war. A single submarine penetrated the Royal Navy's harbour at Scapa Flow and sank the battleship *Royal Oak* while it was at anchor in October 1939. The menace of the U-boats increased after the fall of France in July 1940, for they could now use French ports as bases for attacks on ships far out in the Atlantic.

As protection against U-boat attacks, ships travelled in convoys, guarded by Royal Navy destroyers and corvettes. The Germans soon took to hunting in 'wolf packs' of fifteen to twenty U-boats, which waited in a line across likely convoy routes, ready to attack. By early 1941 so many British ships were being sunk that Prime Minister Winston Churchill started talking about the 'Battle of the Atlantic'.

This battle widened after the United States joined the war as prior to this, U-boats had attacked only British ships. Now they also attacked American shipping on the US east coast, in the Caribbean and then in the Gulf of Mexico, sinking more than 200 ships before the US Navy began its own convoy system. Checked by the convoys, the U-boats moved for a while into the South Atlantic, attacking ships off the African coast taking goods to Britain from the Far East.

The German U-Boat menace - represented here by U-278. *(Martin Bowman Collection)*

By mid-1942, the situation had reached a critical stage and the prospects of the future looked worse. It was obvious to everyone that the Allies could not continue to sustain such heavy losses over an extended period of time. American men and supplies had to be moved over great distances. They would travel over thousands of miles of open seas far out of the reach of air cover and with only a limited number of fighting ships for protection. The small numbers of Navy fighting ships was completely inadequate to handle the U-boat menace.

As the Battle of the Atlantic progressed the American and British Allies benefited from new inventions which helped them to find and destroy U-boats. For example, High Frequency Direction Finding (known as Huff Duff) could detect even very short radio signals from a surfaced U-boat, allowing convoys to be directed away from the area. Of special help to the Allies was the discovery of an 'Enigma' coding machine on board a captured U-boat. This allowed British intelligence workers to decipher the secret Ultra codes which the Germans used for sending orders to the U-boats. Despite all this, the Allies were losing the Battle of the Atlantic.

The bulk of the US Navy's fighting ships were needed to support military operations elsewhere. Allied shipping continued to fall prey to German U-boats that roamed the seas almost at will. In the early days of 1942, the situation seemed almost hopeless. It was clear that the Allies would have to redouble their efforts in the production of military goods to compensate for these staggering losses. They would have to resign themselves to the enormous loss of life and hope they could hold on while limited naval power destroyed the German U-boats one-by-one over a period of years. The big question was could they build enough ships fast enough to hold on?

F H Hoge, Jr. of the War Production Board's planning committee pondered the latest report of losses to German submarines. Making imaginative use of America's industrial and technological strengths to solve such problems was the mission of the Planning Board. On 22 May Hoge submitted a thirteen-page secret memorandum to the chairman of the planning committee of the War Production Board. In it he proposed a '...new method of transportation': not simply conventional cargo aircraft, but flying boats larger than any ever built.

His arguments were logical. Aircraft could leapfrog the submarine menace and make multiple crossings during the time it took a surface ship to make one trip. It was known that aircraft efficiency for load-carrying over long ranges went up with size: he pointed out that existing aircraft devoted thirty-eight percent of takeoff weight for a transoceanic flight to fuel and oil, but that a 300,000 pound aircraft would use only nineteen to twenty percent of the takeoff weight for that purpose. As to landing gear and runway requirements for such large machines, Hoge stressed that flying boats would need neither-this would also save the weight of landing gear, which accounted for fifteen percent of the net weight of conventional aircraft. Finally, suitable harbours existed around the world, even in combat areas where land-based facilities did not exist.

During those dark days of 1942, there was one bright spot for the Allied cause - the ever-increasing number of American Liberty

US industrialist Henry John Kaiser (b. 9 May 1882 – d. 24 August 1967)

One of two
surviving Liberty
ships preserved in
the United States,
Jeremiah O'Brien is
the last un-altered
Liberty. *(Martin
Bowman Collection)*

ships. The Liberty ship was the brainchild of American industrialist Henry
J. Kaiser who had virtually reinvented ship building and placed it on a mass
production basis.

He was born on 9 May 1882 in Sprout Brook, New York and worked as
an apprentice photographer early in life, but was running the studio by the
age of twenty. He used his savings to move to Washington State in 1906
where he started a construction company that fulfilled government contracts.
In 1914 he founded a paving company, one of the first to use heavy
construction machinery. His firm expanded significantly in 1927 when it
received a $20-million contract to build roads in Cuba. In 1931 his firm was
one of the prime contractors in building the Hoover Dam on the Colorado
River, and the Bonneville and Grand Coulee Dams on the Columbia River.

Kaiser's success came from identifying needs and filling them. His
approach to problem solving was to dream big and then to scheme like hell
to make them come true. Amazingly, no matter how grandiose and
impossible his schemes may have first appeared, his aggressive energy,
disciplined use of time, and genius for organisation, inventiveness, and
improvisation had always seen him through to success.

He had never built a ship before, but he set up shipyards in Seattle and
Tacoma, where he began using mass-production techniques that built cargo
ships with an average construction time of 45 days. These vessels became
known as Liberty ships. Though British in conception, they were adapted by
the US as they were cheap and quick to build, and came to symbolise US
wartime industrial output. Based on vessels ordered by Britain to replace
ships torpedoed by German U-boats, they were purchased for the US fleet
and for lend-lease deliveries of war materiel to Britain and to the Soviet
Union via deliveries through Iran. Eighteen American shipyards built 2,710
Liberty ships between 1941 and 1945, easily the largest number of ships
produced to a single design.

Kaiser became world renowned when his teams built a ship in 4 days. The keel for the 10,500 ton *Robert E. Peary* was laid on 8 November 1942, and the ship was launched in California from the Richmond Shipyard Number 2 on 12 November, four days and 15½ hours later. The previous record had been 10 days for the Liberty ship *Joseph M. Teal*.

A visit to a Ford assembly plant by one of his associates led to the decision to use welding instead of traditional riveting for shipbuilding. Welding was advantageous in that it took less strength and it was easier to teach thousands of employees, mostly unskilled labourers - many of them women. Kaiser also adopted the use of sub-assemblies in ship construction; formerly, hundreds of labourers crowded together to complete a ship. Though this practice had been tried on the east coast and in Britain, Kaiser was able to take full advantage of the process by constructing new shipyards with this in mind.

Other Kaiser Shipyards were located in Ryan Point, Vancouver on the Columbia River in Washington state and on Swan Island in Portland, Oregon. A smaller vessel was turned out in 71 hours and 40 minutes from the Vancouver yard on 16 November 1942. The Kaiser hulls also became America's escort carriers, over one hundred small aircraft carriers employed in both the Pacific and the Atlantic theatres. The concepts he developed for the mass production of commercial and military ships remain in use today.

Through his membership in a group called the Six Companies, Kaiser also had a major role in the Joshua Hendy Iron Works of Sunnyvale, California which built the EC-2 triple expansion steam engines for the Liberty ships. Kaiser and his associates organised the California

The Kaiser Permanente Metals Yard No. 1 in Richmond, California. The two large buildings in the upper right corner of the photo are the giant fabricating shop set up halfway between No. 1 and No. 2 yards in which whole upper structures of ships were built in only three units. The fabricating shop served both yards and the fabricated sections were transported to the slipways for installation. *(Russell Plummer Collection)*

Shipbuilding Corporation. and also found time to serve as National Chairman of United Clothing Collection for International War Relief.

Kaiser had justifiably earned a reputation as a doer. When he spoke, which was often, most people listened. Henry J. Kaiser was a loquacious, garrulous individual and the press was always there to listen and report whatever he had to say. It was not only what he said that made headlines, but the manner in which he said it with grandiose use of superlatives and adjectives. He was well used to controversy and his eloquent manner, coupled with constantly outstanding performances, made anything he proposed difficult to discount.

The problem facing the Allies was no different to him than any other problem he had faced in the past. Kaiser's answer was basic; apply simple logic and come up with an elementary solution. *'Stay away from the complex solution as it will generally create more problems than it will solve'* he was supposed to have said, and he was applying that logic to this grave problem which faced the nation.

Kaiser formulated his ideas and prepared to call a news conference on the first of many of his plans to combat the German U-boat menace. He was opposed to the usual approach to things. He detested being caught up in Governmental red tape and had learned to use the press as a means of levering open all the right doors. If he could build public and press support for his ideas, he got more effective results.

Kaiser considered the bureaucracy of Washington DC a disaster and a major obstacle in his way of getting things done. Once he had formulated a plan, he did not want to be hindered by any government's inability to act on anything. He was on record as considering most politicians as masters of indecision. He felt politicians preferred to wait and see if problems would dissolve and not act until there is no other course open to them. He was not a great deal more flattering in his opinion of the military. Right or wrong, those were Henry J. Kaiser's beliefs and his way of doing things - the results

Governor Charles Sprague, Henry Kaiser and President Franklin Delano Roosevelt tour one of Kaiser's Oregon Shipyards, *(Russell Plummer Collection)*

of these beliefs and actions were a vast number of accomplishments and very few failures.

Once summoned, the press gathered to hear what Kaiser had to say. They were aware of how newsworthy he was. They were well aware that not only could he generate the right headlines, but could create the kind of political controversy that sold newspapers. He began by telling the press of the grave problems the Allies faced with the U-boat menace and that he had devised a solution to the problem.

Kaiser chose Sunday 19 July as the day to unveil his proposal. At the launching of the liberty ship *Harvey W. Scott* in Portland, Oregon he said. *'There is no secret concerning the fact that the toll of merchant ships in the western Atlantic since our entry into the war will soon reach the appalling figure of 400. I tell you frankly that this is a matter which has given me many a sleepless night.'*

One of the many small escort carriers as envisaged by Henry Kaiser. This particular example was laid down as the US mercantile *Mormacpenn,* under a Maritime Commission contract (hull number 161). It was acquired by the USN, designated AVG-8 and named *Block Island,* after a sound off the south coast of New England. It was redesignated ACV-8 on 20 August 1942 and then transferred to the Royal Navy and commissioned as HMS *Hunter* (D80). She was originally to have been named *Trailer.* (*R Plummer Collection*)

Then Kaiser outlined his plan. *'Our studies indicate that the answer lies in the aerial freighter... Our engineers have plans on their drafting boards for gigantic flying ships beyond anything Jules Verne could ever have imagined. There are plans for flying ships of 200 tons, and after that plans for ships of 500 tons.'*

As a first step he proposed that selected shipyards on the Pacific, the Gulf, and the Atlantic coasts mass produce seventy-ton Martin Mars flying boats capable of carrying a fourteen-ton payload. *'...that ship would carry a hundred men fully equipped. Five thousand of them could land 500,000 equipped men in England in a single day. And the next day they could fly over again with 70,000 tons of fresh milk, beefsteaks, sugar, and bombs. No submarine could shoot them down.'*

Comments like this were typical Kaiser - his sweeping statements may have provided wonderful sound bites to the press, but for example, the Mars had a maximum speed of 191 knots/221 mph and a crusing speed of 165 knots/190 mph. The transAtlantic crossing with fully equipped troops on board would have taken at least 15 hours in still air - and his comment does not even take into account the return journey before setting out again with beefsteaks and bombs - he was not the sort of man to let details get in the way of good copy!

He proposed that hulls of giant flying boats be built on shipways, launched into the water, and towed to outfitting docks for addition of wings, engines and finishing parts.

This flying boat would have two hulls side-by-side and be able to land and take off in the sea or it could be beached like a landing craft. He told the press he envisaged a fleet of 5,000 such machines. He explained how enemy submarines could have little effect on a cargo ship flying 10,000 feet above the sea. He was prepared to show that this was a serious proposal; he brought sketches and specifications of his twofold proposed plan.

Henry Kaiser's lack of knowledge of aircraft design problems should have been apparent. He eulogised under questions from the press, that he could have such an aircraft designed, built and tested within ten months after receiving the go-ahead. He also boldly stated that he would build five hundred aircraft of this type a year. The press loved it and Kaiser's plan made the front pages. It was kept there by the continuing release of further details at subsequent news conferences.

Kaiser's plan for the small 'escort' aircraft carriers was taken more seriously and was investigated more rapidly than the flying cargo ship idea. The Navy's high command reviewed Kaiser's plan for small, fast anti-submarine aircraft carriers built on merchant ship hulls. According to Kaiser, they turned him down in a 16:0 vote without explanation. Kaiser then approached the Secretary of the Navy Forrestal without success.

The first unofficial Navy reaction was noncommittal. When asked by the press for any comment on Kaiser's proposal, Adm. Howard L. Vickery, vice chairman of the United States Maritime Commission and head of the wartime merchant ship programme, who was on hand to present Kaiser's Oregon Shipbuilding Corporation with a gold star for outstanding production achievement, commented, *'I am a shipbuilder, not an airplane builder.'*

Kaiser followed up with a series of dramatic and highly publicised press statements. Because of his proven track record as a miracle worker the general public and many members of Congress were prepared to believe his most extravagant claims. When he pledged that *'...with the aid of the aviation industry and with the equipment already in place in the shipyards we can have the assembly line in production at six months or less'* people listened. Kaiser also found sympathetic ears in the Roosevelt administration as one of the few industrialists who had supported the president.

On 29 July Capt. Eddie Rickenbacker, the World War One flying ace and president of Eastern Airlines, whose aviation judgment commanded enormous respect, testified before the Senate committee considering Kaiser's proposal. Rickenbacker did not see how Kaiser could make good on his claims. *'There's a hell of a lot of difference between building ships and making airplanes'*.

This stung Kaiser. The following day he testified before two Senate committees, one investigating cargo flying boat prospects and the other considering legislation to promote such planes. Kaiser said that until 1940 he had never even seen a ship launched and that *'Army engineers told me it would be impossible to build Bonneville Dam. It doesn't matter what can't be done...'* he said. *'...so long as we have to do it, we can do it.'*

Not until Kaiser's intensive publicity and lobbying campaign was well under way did he go to the Army and Navy with his proposal. Responsible military officials thought the proposal impracticable, but they knew that the public and certain members of Congress believed otherwise. Not wanting to give the temperamental industrialist an unqualified no, they asked the Aircraft Division of the War Production Board to review his proposal and give him a complete insider's picture of production problems, shortages, schedules, and objectives.

Kaiser felt he had played the game long enough and went directly to the White House to see the President. He spoke to Marvin H. McIntyre who was chief secretary to President Roosevelt, McIntyre told him, *'I don't think you can see the President, but if you write a letter of no more than four paragraphs, I'll lay it on the President's desk and see that he reads it.'*

Kaiser wrote the President a one-page letter explaining his plan for the escort aircraft carriers and their proposed use. He told him in the letter of the Navy's lack of interest in the plan. The letter was, as promised, laid on the President's desk and was read.

Shortly afterwards, Kaiser received a phone call from Rear Admiral Emory Land, chairman of the maritime commission who supposedly shouted at Kaiser, *'What the hell are you doing? Come to my office in the morning.'* Land was charged with overseeing the design and construction of the more than 4,000 Liberty ships and Victory ships that flew the U.S. flag during World War Two and concurrently served as Administrator of the War Shipping Administration. Thus Admiral Land exercised authority over both construction and allocation of non-combatant maritime assets to Army, Navy and commerce. Kaiser was thus satisfied that he was about to get serious consideration for part one of his plan.

In the meeting the next day, Kaiser's shipyards received an order for 100 baby aircraft carriers. Then it seemed that all the brass in the United States Navy came down on Kaiser and his carrier plan. The Navy wanted destroyers, not aircraft carriers, and wanted them now. Henry explained that he couldn't build destroyers. As a result, his order for the 100 carriers was reduced to 50.

Kaiser started to build the carriers and they were highly successful. The

Vice Admiral Emory Scott Land (*b.* 8 January 1879 – *d.* 27 November 1971, chairman of the US Maritime Commission.

Navy changed its mind again and came to Kaiser desperate for more of the Baby Flat Tops as they became known by the US Navy or 'Woolworth carriers' by the British Royal Navy. Kaiser's shipyards turned them out in record numbers, as many as six a month, far exceeding the expectation of everyone - including Henry J. Kaiser.

Kaiser's plan for the flying cargo ship was falling prey to the military who doubted that such a machine could fly and they knew that Kaiser had no experience in either designing or building aircraft. The military believed that to entrust such a project, even if they believed in it, to a man or company with no aviation experience would be foolhardy. Kaiser was not so easily dissuaded and continued his publicity campaign to gain public and press support for the flying cargo aircraft project. Meanwhile, he was busy contacting and seeking an alliance with some of the major aircraft companies in the country - Martin, Douglas and Northrop.

In July 1942, Kaiser went to Glen L. Martin, one of the pioneers of American aviation, with a proposal for a six-company combine to design and construct the 5,000 flying cargo aircraft. Kaiser was convinced that if he could show the backing of six major aircraft companies, all combined with his production know-how, the government would have to listen. Martin turned him down, as did every other aircraft company he contacted. They all felt the project was unrealistic. Not one felt that anything as huge as Kaiser had proposed could possibly fly and that the capabilities Kaiser spoke of were beyond the technology of the day.

Kaiser continued to push for public and government support, but without success. Kaiser went back to see the President just as he had about the escort carrier project. This time McIntyre told Kaiser, *'I don't think you'll get any consideration'*. He left without writing a note or seeing the President.

Merril C. Meigs, of the War Production Board, said later that what Kaiser didn't know was that the public pressure was having an effect on the people at the War Production Board and other government agencies. They were having concerns about the public support Kaiser was receiving for his programme. They were fearful the public would believe that the government was passing up a chance to overcome the U-boat menace and speed the success of the war.

Merrill Church Meigs (*b.* November 25, 1883 - *d.* January 26, 1968)

Kaiser's continued pounding of the government for not acting on his proposal was intensified by his statement, *'I don't care if I don't make a dime out of this deal, I'm motivated purely by patriotism.'* This statement brought increased pressure from the press for the government to respond to Kaiser's offer or at least reply with other than their standard answer of *'The whole scheme is utterly fantastic.'*

Finally, in early August Nelson, ordered Jesse H. Jones of the Reconstruction Finance Corporation to open contract negotiations with Henry J. Kaiser on his proposed 5,000 flying cargo ships and to issue a letter of intent. Jones, however, insisted that Kaiser stick to his claim of patriotism and ordered the contract to be one without profit.

The Reconstruction Finance Corporation (RFC) was an independent agency of the United States government, established and chartered by the United States Congress in 1932 during the administration of President Herbert Hoover. When Eugene Meyer became Governor of the Federal Reserve Board, he had suggested creating the RFC. The goal of the RFC was to boost the country's confidence and help banks return to performing daily functions.

Even before World War Two began the RFC's power were expanded and further expanded. President Roosevelt merged the RFC and the Federal Deposit Insurance Corporation (FDIC), which was one of the landmarks of the New Deal. Oscar Cox, a prime author of the Lend-Lease Act, general counsel of the Foreign Economic Administration joined as well. Lauchlin Currie, formerly of the Federal Reserve Board staff, was the deputy administrator to Leo Crowley.

Donald Marr Nelson (*b*. 1888 – *d*. 1959) Chairman of the War Production Board.

From 1941 through 1945, the RFC authorised over $2 billion of loans and investments each year, with a peak of over $6 billion authorised in 1943. The magnitude of RFC lending had increased substantially during the war.

Kaiser needed an alliance with someone in the aircraft business in order to succeed. None of his companies had the kind of engineering talent that the design and prototyping of the giant flying cargo ships would require. There was an agreement not to raid the engineering staffs of the other major aircraft companies.

When one looks deeply into what Kaiser was proposing, his plans for the flying ship were nebulous, at best, and without any engineering to back them. What he held in his hand were nothing more than artists' concepts and a specification sheet of what was thought to be desirable in characteristics, performance and capabilities.

Kaiser also seemed to be stymied at every turn by officialdom. He had no co-operation from other aircraft companies and he was told that he would not be allowed to employ aircraft engineers or be allowed allocations of strategic materials such as aluminium.

In fact, when one studies the press material, it becomes clear that during July and August 1942, Kaiser actually made two proposals, although they were not always clearly differentiated. One was for the mass production of Martin Mars-type flying boats in association with shipbuilders and the aircraft industry. (Kaiser claimed that the Mars had been recommended to him by Donald Douglas.) The second proposal was for the design and

Aviation pioneer Grover Cleveland Loening (*b*.September 12, 1888 – *d*. February 29, 1976), advisor to Donald Nelson.

development of a 200-ton flying boat. Among the agency heads to whom he made these proposals was Donald Nelson of the War Production Board. This was a wartime agency of huge power and responsibility. Its head, the former Sears president Donald Marr Nelson, was a Roosevelt man and sensitive to the opinions of both his boss and the public. He avoided making hard decisions by finding ways to postpone them or to compromise.

Advising Nelson on aviation matters was Grover Loening, who at Columbia University in 1910 received the first master's degree in aeronautics in the United States before becoming assistant to Orville Wright. During The First World War Loening headed the Loening Aeronautical Engineering Corporation and afterward pioneered in the development of a variety of military and civilian aircraft including the Loening Amphibian. Loening had a soft spot

for seaplanes. He was a wealthy man and had served on the board of directors of Pan American Airways in the late 1930s during arguments as to whether landplanes or flying boats should be used on the Atlantic runs. Loening's argument was that putting money into landplanes was foolishness, and that you shouldn't attempt to fly the Atlantic except in big flying boats.

The position of the War Department had become known earlier that summer when reporters talked to Under Secretary of War Robert Patterson outside the hearing when Kaiser testified on July 30th. According to Patterson, the production of cargo planes would not be allowed to impede production of fighting planes, *'...as we must supply not only ourselves but our allies'*

Aircraft manufacturers and high military and civilian officials communicated their concern to Senators Wallgren, Hatch, and Burton, who sat on the Subcommittee on Aviation and Light Metals of the Senate Special Committee to Investigate the National Defense Program. The senators made their points to the chairman and the committee in an August 5, 1942 memo:

1. There was already in effect a 'well developed air cargo program' that included transport planes of the Army and Navy. The cargo ship program included the Lockheed Constellation and the Martin Mars.

2. There was a tremendously important element of time involved in the production of the Kaiser Flying Cargo Ships. It is generally estimated that it would take two years to get into any appreciable production of the flying ships and that they still would be experimental models.

3. Recognizing that the shipbuilding and present airplane progress cannot be permitted to cease it is obvious that the addition of a great program of Flying Cargo Ships would aggravate many of the existing transportation, labor and housing shortages.

4. In view of the extraordinary genius and demonstrated production record of Henry J. Kaiser, any suggestion from him is entitled to respectful and careful consideration. On the other hand when the proposal involves entering into a new industry and possibly jeopardizing the entire war program of the country, it must also receive a thorough analysis and evaluation of its practicality before it is endorsed or undertaken.'

5. The Subcommittee discussed this matter confidentially with executives of the aviation industry. Each of them indicated a generous and patriotic willingness to cooperate in every way possible with this or any other development in the aviation industry. There was, however, a striking unanimity of opinion emphasizing the probable impracticality of the Kaiser proposal in the form in which it was presented through the newspapers.

On Friday afternoon, 7 August, Kaiser and his executive assistant, Chad Calhoun, were ushered into Donald Nelson's office in the Social Security Building. According to Calhoun's later testimony, Nelson's first comment was, *'...Henry, I am going to take another chance on you.'* This referred to Nelson's previous authorisation for building Kaiser's Fontana, California steel plant over strong opposition. Then, according to Calhoun, Nelson showed them a letter of intent prepared for the signature of Admiral Towers that would authorise Kaiser to proceed with the construction of 500 Mars-type flying boats. Later, the Navy balked at this and Towers never signed the letter.

Kaiser read the draft and then said, *'How about the letter of intent to proceed with the design of the 200-ton plane?'* Nelson said that that would probably come through also, but probably not until after he returned, as he was leaving on a trip that weekend.

However, on Monday 10 August while Nelson was away, William Batt and Merrill Meigs of the War Production Board gave Kaiser two letters signed by Donald Nelson which Calhoun said *'...kissed off both ideas'* by giving Kaiser a series of *'...impossible conditions.'* Kaiser later said that he thought that Meigs and Batt had waited until Nelson was out of town and then given him letters with rubber-stamped signatures.

Chief of the Navy's Bureau of Aeronautics Admiral John Towers later testified before the Senate Committee during the 1947 hearings that: *'...Based on my experience over a long period of years in the procurement of aircraft and on my recent experience in connection with our contract for one Martin Mars, I was confident that it would be impossible for Mr. Kaiser to come anywhere near approaching his predictions. I felt that the development of plans and the construction of such planes in quantity could not be accomplished in time to be of any aid and assistance during the war.'*

Admiral John Henry Towers (*b*. January 30, 1885 – *d*. April 30, 1955)

Turning The Corner...

Just when the whole project seemed in doubt, Kaiser received a ray of hope. Kaiser had been told that Howard Hughes was really his best bet to get the flying cargo ship designed, that Hughes was the finest designer of special purpose aircraft in the world, and that Hughes was also one of the most gifted innovators in American aviation.

Hughes' record of accomplishments impressed Kaiser and he was happy to learn that Hughes might be available, especially when he heard that Hughes had a score to settle of his own with the brass in Washington for denying him a major aircraft contract. This information was all Kaiser needed to know and he began tracking down the elusive Hughes.

Late in the summer of 1942, Kaiser learned that Hughes was in the St. Francis Hotel in San Francisco under an assumed name.

On Thursday Kaiser departed Washington for New York and on Sunday boarded a train for Oakland and a meeting with Howard Robard Hughes.

HRH

Just like his entire life, the early days of Hughes' life are full of legends, supposed quotes and unatrributable stories about him and his father. Likewise there are many factual mysteries and discrepencies. Most are impossible to corroborate and ascribe to primary souce documentation. That said, they are certainly worth inclusion here in with the known facts, for they do create an overall picture of the forces that shaped his family and his life.

Hughes's father, known by many as 'Big Howard' after young Howard's birth, was born in 1869. The son of a Lancaster, Missouri lawyer who later moved his family to Keokuk, Iowa, Big Howard graduated from Harvard, earned his law degree at the University of Iowa, then worked alongside his father.

But the private practice of law in his father's shadow soon proved to be not the life for Big Howard. By all accounts he yearned for a more risky, chance-taking life, and the stories of big mineral strikes in neighboring states captured his imagination. He sought to strike it rich in lead or zinc in nearby Missouri. Then the news of the great Spindletop oil strike lured him further south to Texas.

With his legal background, Big Howard made a good 'lease hound'; that is someone in the petroleum industry who negotiates with land-owners for options, oil-drilling leases and royalties along with negotiating with producers for the pooling of production in a specific oilfield. He came in contact with Walter Benona Sharp and together they formed a partnership with a drilling company that probed for oil for themselves or under contract for others.

Big Howard's fortunes fluctuated wildly. He was a classic entrepreneur, trying and failing at many endeavors before he finally found his speciality. Being a lease hound and a wildcatter were not the easiest ways to make a living; a few struck it rich, but many more failed. When Big Howard was up, he spent it. And when he was down, he was usually deeply in debt. During one of his up periods he married Allene Gano, a beautiful young woman from a well-known Dallas family, on 24 May 1904 and the couple left for an extended honeynoon to England, France and Germany that was reported as a 'world tour'. She was the daughter of William Bariah Gano, a descendant of Catherine of Valois, Dowager Queen of England, by second husband Owen Tudor, and wife Jeannette de la Fayette Grissom. When they returned to the Hughes' home at 1404 Crawford Street, Houston, Texas he was nearly penniless and she was pregnant.

Howard Robard Hughes, Snr. *b.* 9 September 1869 - *d.* January 14, 1924)

Thus was born Howard Robard Hughes, Jr. His birthplace is recorded by historians as either Humble or Houston, Texas. The date is also uncertain, though Hughes claimed his birthday was Christmas Eve. A 1941 affidavit birth certificate of Hughes signed by his aunt Annette Gano Lummis and Estelle Boughton Sharp states he was born on 24 December 1905, in Harris County, Texas. No time of birth is listed. On Record No. 234358, of 29 December

1941, that was filed 5 January 1942, with the Bureau of Vital Statistics of the Texas Department of Health. However, his baptismal record of 7 October 1906, in the parish register of St. John's Episcopal Church, in Keokuk, Iowa, has his birth listed as 24 September 1905, without reference to the place of birth. The handwriting of the baptismal record is a rather trembling one, which has led to speculation that the clerk was an aged person and there is a good chance that, supposedly, being hard of hearing they misheard 'December 24' as 'September 24' instead.

Howard Jr in his childhood.

When 'Little Howard' or 'Sonny' as he was sometimes called, was just over a year old, the standard fishtail bits on one of his father's drilling rigs began rapidly blunting as they hit hard rock at a site near Pierce Junction, Texas. The same frustrating experience was repeated at Goose Creek. The problem was that the fishtail shape of the standard bit had only two cutting edges whose scraping action was nearly useless against rock. The partners agreed that there had to be a better way.

Without doubt Big Howard was an inventive and innovative tinkerer, and he supposedly lucked into an inspired idea. Within the Texan oil industry, legend has it that Big Howard met a young millwright in a bar who was down on his luck. The young man showed him a model of a crudely fashioned drill made of wooden spools. Big Howard bought it on the spot for $150!

Howard with his mother Allene (*b*. 14 July 1888 - *d*. 29 March 1922.)

Prior to 1909 the traditional fishtail bit scraped the rock and quickly dulled in service. The Hughes two-cone bit's revolutionary rolling action crushed hard-rock formations with twin cone-shaped, hardened steel bits, each with 166 cutting edges, revolving on bronze bearings shaped to provide a large surface with reduced friction.

Unlike the fishtail bit, the Hughes two-cone bit rolled in a true circle, crushing and grinding the rock. This rolling motion allowed the cutting edges to chip the rock, one edge after another. Significantly, the cutting edges were designed so they would avoid falling into previous cuts, preventing what is known as tracking. This enabled each edge to continuously crush a new portion of the rock. The absence of scraping with the fishtail bit allowed the Hughes Two-Cone Drill Bit to drill faster and further

A group of Texas wildcatters with seven 'fishtail' drilling bits of assorted sizes that pre-dated the two-cone Hughes design.

HOTEL HUGO
SOHO

jamanda@voyage prive . com

#6 N/8/26

525 Greenwich Street, New York, NY 10013
P: 212.608.4848 F: 212.608.4844
www.hotelhugony.com

The US Patent drawing of the Hughes Two Cone Drill Bit, that states that Howard R. Hughes was 'the inventor'. This was certainly not the case.

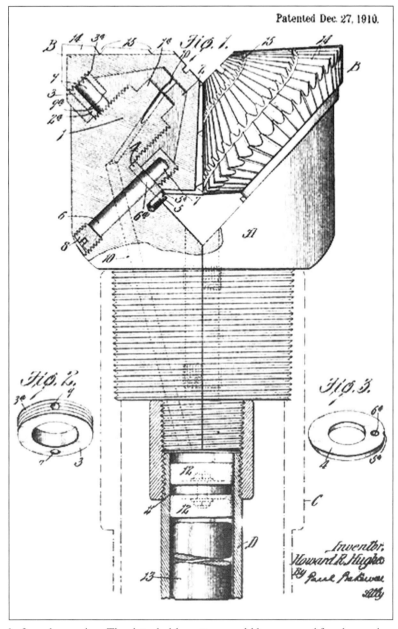

before sharpening. The detachable cutters could be removed for sharpening or replacement when worn.

The bottom of the drill hole, as formed by the operation of the bit, was a perfect seat for a water-tight joint, preventing leakage after the casing had been set. The Hughes Two-Cone Drill Bit ushered in a new era of abundant, inexpensive fuel and laid the foundation for the automobile age.

On 20 November 1908, Big Howard filed the basic patents for the Sharp-Hughes Rock Bit, and on 10 August 1909 was granted two U.S. Patents 930,758 and 930,759 for this rock drill. Hughes' two-cone rotary drill bit

A group of oilmen with a Sharp-Hughes two-cone drilling bit - it was this device that was to bankroll Hughes Jr's aviation activities.

penetrated medium and hard rock with ten times the speed of any former bit, and its development revolutionised oil well drilling. He may not have actually invented the two-cone roller bit, but his legal experience helped him in understanding that its patents were important for capitalising on the invention.

Hughes shrewdly went on to patent the idea in every major patent office in the world, thus achieving a monopoly. As a final strategy that would assure Howard Jr. of a virtually unlimited money supply, they established the policy of leasing the bit at a fixed price whether oil was struck or not. Bits were never sold outright. When they became worn they had to be returned to the Sharp-Hughes Tool Company for overhaul and refurbishment.

Years later a reporter is supposed to have asked Howard Hughes, Jr. whether it was true that the Hughes Tool Company was a monopoly. The legend has it that Howard Jr replied *'Of course not. People who want to drill for oil and not use the Hughes bit can always use a pick and shovel.'*

When drillers discovered just how efficiently the Hughes bit cut through rock, the demand was instantaneous and enormous. Walter Sharp died in 1912, before the company became truly prosperous, and his widow soon discovered that Big Howard often spent more than the company earned. Fearing that his

The original Sharp-Hughes Tool Co. building in Houston.

Howard Hughes with his self-made motorcycle.

'business expenses' would bankrupt the company, and her with it, she sold her half interest in the company to oilman Ed Prather, a friend of both families. Big Howard's spending habits continued uncontrolled, and after three years of worry Prather sold out to Hughes. So the one-shareholder Hughes Tool Company came into being, the golden wellspring of his son's future wealth.

During his formative years young Howard led an affluent but lonely life. Big Howard continued to work and play hard. He and Allene were an extrovert, popular couple and young Howard was frequently left alone.

Howard envied those who, like his father, mingled easily and well. Hughes Jr supposedly said when discussing this: *'My father was plenty tough. He never suggested that I do something; he just told me. He shoved things down my throat, and I had to like it. But he had a hail fellow-well-met quality that I never had. He was a terrifically loved man. I am not. I don't have the ability to win people the way he did. I suppose I'm not like other men. Most of them like to study people. I'm not nearly as interested in people as I should be, I guess. What I am tremendously interested in is science, nature in its various manifestations, the earth and the minerals that come out of it.'*

Howard Jr showed a mechanical and technical aptitude from an early age. At the age of 12 he motorised his bicycle by installing a storage battery and an automobile starter motor which he bought at a junkyard, and there is certainly photographic evidence of this. Other sources state that he once built a motorcycle from parts taken from his father's steam engine. Later he built a radio transmitter and receiver, and Howard obtained one of the early ham radio licenses, call-sign SCY. Big Howard encouraged his son's technical interests and provided a worktable at the plant where he could learn about machines and work with tools.

Howard's mother worried excessively about his health and went to great lengths to protect him from exposure to illness, and many writers have alleged it from her that Hughes Jr acquired his fear of germs. Certainly there is enough evidence to suggest that his mother may have suffered from a form of what is today known as Obsessive-Compulsive Disorder (OCD), and coddled and spoiled her only child. It was Hughes' mother who first provided her young son with a means of escaping social situations and pressures by using the excuse of illness, termed as Social Avoidance Behaviour. As a young boy, when Howard wanted to attend summer camp during a time when the public feared the spread of polio, his parents naturally wanted assurances that their son was protected. When this assurance was not forthcoming, his mother decided it was better to keep him home. Subsequently after attending camp one summer, Hughes avoided another year at camp by complaining about headaches and bad dreams when he returned home. Later, on the verge of adolescence, young Howard became ill and was kept out of school for most of the year. He developed a form of paralysis that was never diagnosed and which disappeared after several months.

Looking very smart, and really quite 'modern' - a picture of the teenage Howard Hughes.

Hughes's First Flight

In the autumn of 1920 when he was fourteen years old, Howard Jr was sent to the Fessenden School at West Newton, Massachusetts. There he flunked spelling, got one hundred percent on the algebra winter term final exam, did well in other subjects, and finished the year in the top third of his class. He played sax with the school jazz band, was a bench-warmer on the senior football team, runner-up in the golf tournament finals, and wrote for the school magazine. He was remembered by teachers and fellow students as intelligent, quiet, shy and retiring, but witty when he did join in conversation.

1920 was also the year of Howard's first flight. He and his father had attended the Harvard-Yale boat races, and in the flush of his alma mater's victory Big Howard took his son for a ride in a Curtiss flying boat that was hopping passengers at five dollars a head from the Thames River near New London, Connecticut. Like so many of this era, this short trip sparked off a life-long facination with flying.

While young Howard was at Fessenden some of Big Howard's oil business shifted to Southern California, where his younger brother Rupert worked for Samuel Goldwyn writing scripts for the movie industry. In a fateful move, young Howard was transferred to the Thatcher School in Ojai, California, fifty miles from Hollywood. Subsequent visits to movie sets with his uncle Rupert sparked his two other interests: moviemaking and beautiful girls. These were interests that would prove enduring.

Then tragedy struck. During the spring term following his arrival at Thatcher, when Howard was sixteen, his mother died undergoing surgery for an entopic pregnancy in a Houston hospital during March 1922. After the funeral his father persuaded Annette Gano, his wife's younger sister, to look after him for a year or so. Midway through the school year Howard dropped out of Thatcher and he and Annette lived in Pasadena. His father somehow managed to have Howard Jr admitted to classes at the California Institute of Technology (commonly referred to as Caltech), a private research university located in Pasadena, California. Caltech had a strong emphases on science and engineering. This Big Howard did by contributing to the scholarship fund, even though young Howard had not yet finished high school.

Meanwhile he had applied for admission and was admitted into the Rice Institute in Houston. He did well there, especially in math. But at the end of his first term, just as he turned eighteen, tragedy struck again, for his father died of a sudden heart attack.

Big Howard had prepared a new will after Allene died, but had never signed it. Thus the will still in effect, then over ten years old, left half of the estate to Allene, twenty- five percent to young Howard, and the rest to his grandparents and his uncle Rupert. Under Texas law Howard received his dead mother's share of the company, but he wanted full control of Hughes Tool and tried to buy out his relatives. There were two problems: they were reluctant to sell and he, at eighteen, was not legally competent to take over. Instead he went to Europe. By the time he came back he had decided on what to do.

The Texas Civil Code provides that if a minor can convince the court that he has the ability to handle his own affairs he can be declared competent to enter into binding contracts. The young Hughes went to court and convinced the judge he was perfectly capable of doing just that. Armed with his new

Ella Botts Rice - Mrs
Howard Hughes.
(*b.* 11 July 1904, *d.* 23
June 1992)

legal status, he employed a tactic that he would use many times in later life – he simply wore his relatives down until they agreed to sell.

So, aged just nineteen, Howard Hughes became sole owner of the Hughes Tool Company. He moved into his father's office and took over, learning the business as he went along. However, he had no lasting interest in his father's business, except for the money that came from it. Like Big Howard, he wanted to get out from under his father's shadow. His visits to Sam Goldwyn's movie sets with Uncle Rupert had exposed him to the glamour, excitement, and opportunities of Hollywood. The motion picture world combined technological challenge with the promise of big rewards just as the oilfields had for his father.

His relatives urged him to complete his education, but instead, on 1 June 1925, he married Ella Botts Rice, daughter of the founding family of Rice Institute. Shortly afterwards the newlyweds moved to Hollywood and Hughes was never to make Houston home again. For a time Hughes and Ella stayed with his uncle Rupert. Then the young couple moved into a suite at the Ambassador Hotel.

In a fortunate decision that autumn, he hired able, experienced Noah Dietrich to manage his business affairs, freeing himself to try to make movies. Dietrich was born in Madison, Wisconsin, to Lutheran minister John Dietrich and the former Sarah Peters. Trained as an accountant, in 1910 he started in business in Maxwell, New Mexico, and later moved to Los Angeles and New York City before moving back to Los Angeles.

In 1925, at the age of 36, Dietrich met 19-year-old Hughes. Dietrich began by running the tool company for Hughes, allowing him to pursue his interests in the film and aircraft businesses. In his memoir, Dietrich observed that Hughes had little interest in ToolCo beyond its being a source of wealth and investment. *'The tool company...'* Hughes is supposed to have said *'...was my father's success. And it always will be.'*

Dietrich would advise and guide Howard Hughes and be his right-hand man for the next 30 years. Dietrich became Hughes' most indispensable executive - 'Noah can do it' was, according to Dietrich's memoir, a frequent Hughes expression whenever difficult, if not impossible, wants or needs had to be met. Some - such as when Dietrich arranged a stock ticker to be installed in a Hughes home - were merely difficult. Others - such as the time Dietrich arranged the shipment of Hughes' large private liquor stock from Texas to California during Prohibition - put him at serious risk.

Noah Dietrich (*b.* 28
February 1889 *d.* 15
February 1982)

Some of Dietrich's duties got him caught in the middle of disputes with members of Hughes' family. During a period when Hughes refused to talk to anyone outside a few business associates, Dietrich recalled, a Hughes aunt accused Dietrich - who turned out to have been kept in the dark about Hughes' exact whereabouts himself - of covering up Hughes' death in order to claim the Hughes empire for himself. When Hughes returned, Dietrich wrote, he made a special point of calling this aunt. *'I didn't want her to continue thinking I was taking over the Hughes empire while keeping her nephew in the deep freeze.'*

As the Hughes fortune began to grow, so did his imagination and dreams. Like many dreamers of big things, he would have dreams that would go sour and end in failure. One dream that went sour was the ill-fated Hughes steam car. He had seen in steam a viable alternative to the gasoline-powered car. Steam had been used to power cars for many years, but it had then all but disappeared from the motoring scene. When compared to gasoline-powered cars, steam cars seemed terribly inferior. The gas engine cars could be started and driven almost immediately. Steam cars took 30 minutes to an hour of required warm-up time. Modern gasoline-powered cars could travel 300 to 400 hundred miles without stopping for fuel. Steam-powered automobiles required a water stop every 50 to 100 miles.

Hughes was convinced that these problems could be overcome with sound engineering. So he set himself the task of creating and manufacturing a revolutionary new automobile, the Hughes Steamer, but where to start? Like everything connected with Howard Hughes, there are doubts and mysteries. Some sources claim he went directly to Dr. Robert Millikan, the famed physicist who had won the Nobel Prize a few years before and president of the California Institute of Technology in Pasadena.

Dr. Millikan considered the problem and referred him to a couple of recent Caltech graduates, Howard Lewis and Bruce Burns. Other sources say that Millikan had nothing to do with it, and that someone in the Hughes organisation contacted Robert L Daughterty, the head of the mechanical engineering department at Caltech.

What is beyond doubt is that Hughes contacted the two engineers, and they were willing to work for him, especially when he mentioned the handsome salaries he was willing to pay and they put together a team of engineers and mechanics to design and build a steam car to his exacting specifications.

The main points were that the car could be started and driven within one minute and that it would be able to travel 350 to 400 miles without stopping for water or fuel. His engineers studied every steam car and machine ever built in an effort to meet the specifications.

The project took from 1927 until 1930. When completed, Hughes came to its unveiling with his right-hand man, Noah Dietrich. When told that the car had met all of his specifications, Hughes was interested in how they had accomplished the major goal of driving 350 to 400 miles without stopping. His engineers told him that the entire body, including the doors, were filled with steam condensers. as water condensation was the key behind making it all work. Hughes was horrified and expressed the fear that even in a minor collision he might be scalded to death because he was sitting in the middle of a giant steam kettle. He ordered the car to be cut up into small pieces.

But this is in doubt, for parts of the car are said to have survived to this day, and later Daughterty went on to state that '...*stories talk about a radiator system that had tubes lining the entire car. This is nonsense. The radiator and boiler system were in the front of every version I saw, so there would have been no danger of being scalded to death. I don't think Hughes saw a complete prototype. Rather, he saw a standard Stanley or Doble steamer with a lot of the components added*'.

The Hughes fortune continued to grow throughout the 1930s at an amazing rate. The worldwide depression somehow seemed to avoid the Hughes enterprises. His great love affair with flying and aviation led him to the inevitable - he would buy an airline. He began buying Trans Western

Airlines stock. By 1940, he had acquired 75 % of it at a cost of $80 million. TWA, as it was known, was one of the major American airlines and Howard Hughes was now in complete control of it.

Shortly after arriving in California in 1926, Howard Hughes purchased 51 percent of the stock of Multi-Color, Inc. This corporation purportedly owned the patents and had the know-how to make colour movies. The far-seeing twenty-year-old believed that some day most, if not all, movies would be filmed in colour.

Hughes was ahead of the times with Multicolor, which was a subtractive natural color process for motion pictures. Multicolor, introduced to the motion picture industry in 1929, was based on the earlier Prizma Color process, and was the forerunner of Cinecolor. The idea was excellent in theory, but in practice it failed to work. He poured money into research, but the experimenters could not evolve a practical process for the development of motion picture colour film.

Multicolor enjoyed brief success in early sound pictures. A number of features included sequences in Multicolor: *This Thing Called Love* (1929), *His First Command* (1929), *Red Hot Rhythm* (1929), *Sunny Side Up* (1929), *Married In Hollywood* (1929), *Fox Movietone Follies of 1929* (1929), *The Great Gabbo* (1929), *New Movietone Follies of 1930* (1930) and *Delicious* (1931). Hughes himself used Multicolor for a sequence in *Hell's Angels* (1930) but printed by Technicolor, as Multicolor could not yet supply as large a demand of printings in such a short amount of time. Multicolor was also utilised in several cartoons of the era. Ultimately, Multicolor turned out to be one of Hughes' few financial flops.

7000 Romaine Street Los Angeles, once the headquarters of Multicolor Inc, it was to become Howard Hughes' base for many years.

Although the project itself failed, out of it Hughes acquired the building at 7000 Romaine Street in Los Angeles, which was used as his headquarters until 1953. At that time he sold the building to Eastman Kodak but apparently later regretted his action, so in 1957 he repurchased it, and used the building as his major headquarters, message centre, and general command post.

Into The Movies

Howards first motion picture *Swell Hogan* by director Ralph Graves was a
flop and never released. Doubling the stakes, he tried again. This time he
hired an experienced director, kept hands off, and achieved a modest success
with *Everybody's Acting* (1927). His third picture, *Two Arabian Knights,*
won an Oscar in 1928 for director Lewis Milestone as for Best Director of
a comedy picture. He was now ready for what he really wanted to do: *Hell's
Angels,* a World War One flying picture.

This time Hughes went all out. He used people and money lavishly,
employed thousands of extras for battle scenes instead of using old newsreel
footage. When the script called for the destruction of a Zeppelin, he bought
the real thing and burned it. He reworked whole flying sequences because
they lacked cloud backdrops and the dogfights appeared static.

He could not find a real German Gotha bomber for the air raid
sequences, he had to settle for Roscoe Turner's twin-engine Sikorsky, at
that time supposedly the largest aircraft in the United States. Hughes's
technicians made the aircraft over to look like a Gotha. Turner flew this
'Gotha' during filming. Turner was a celebrity aviator and as famed for
his natty uniforms and mustache as for his innovative and enterprising use
of aircraft.

Turner was born in Corinth, Mississippi, the eldest son of a poor but
respectable farmer. When America entered World War I, he applied for pilot
training but was turned down because he did not have a college education.
As the need for pilots grew, the education requirements were lowered and
he was trained to be a balloon observer. Privately, however, he was able to
receive aircraft pilot training. He was discharged as a First Lieutenant in
1919. With his discharge payment, he purchased a surplus aircraft and spent
the 1920s barnstorming.

The first half of the 1920s saw Turner living the gypsy life; teaming up
with other ex-Army fliers to perform shows across the USA. He saw there

Stunt pilot Harry Parry
details some of the
aerial manoeuvres for
director Hughes during
the making of *Hell's
Angels.* At one time it
was alleged that the
movie has over 20,000
actors, extras and crew
members on the
payroll.

Opposite page: One of
the publicity posters for
the movie that was to
propel Jean Harlowe to
stardom.

was no future in barnstorming and realised that his fame was in the west. It was during this time he became notable for his Nevada Airlines, flying wealthy people to Reno, Nevada. He was granted the title of Colonel in the Nevada National Guard by the Governor of Nevada for his efforts. This title he kept proudly until his death. He also worked for a time as a stunt pilot for the movies

During one sequence Turner became annoyed with Hughes's demands. *'Hughes was kind of working with his director. He kept wanting me to get lower and lower, so I just put the wheels down and hit the cameras on one trip.'*

Hughes permitted himself another cinematic 'trick': he had all the German aircraft painted black and the allied machines painted white so the audience could clearly tell who was who in the dog-fights.

Hughes's quest for perfection in filming *Hell's Angels* led to an enormous number of re-takes. Normally a director might shoot 80,000 feet of film to produce a ninety-minute film of 8,000 feet. Hughes supposedly shot more than 2,250,000 feet of film- almost 300 times the amount required for the final picture.

Moreover, the film was silent; and by the March 1929 sneak preview movies had become 'talkies'. Overnight the market for silent films had disappeared. Planes zoomed, cannon fired, people talked and shouted, blazing planes smote the earth-but never a sound was heard. So the story was rewritten for the ear as well as the eye and for this second version Howard replaced Grethe Rüzt-Nissen with an English-speaking platinum blonde, Jean Harlow. As a result, *Hell's Angels* took over three years to make and cost $3.8 million at a time when a million dollars was a lot of money.

In the midst of his concentration on movie making Hughes neglected his wife. Ella left to rejoin family and friends in Houston and divorced her obsessed husband in 1929 after four years of marriage. Hughes supposedly locked her bedroom door in their 211 Muirfield home and left everything just as she had left it until the house was sold years later.

The sucess of *Hell's Angels* encouraged Hughes, who then made three flops in a row before making *Scarface* and *The Front Page*. Both of these films received rave reviews and were box office successes. Hughes then decided to rest on his movie laurels for a time and give full attention to flying.

The beginnings of an aviation career.

Hell's Angels also marked the real start of Hughes's flying career. Shooting on the film started in 1927, and in the autumn of that year Hughes began to learn to fly in a serious way at Santa Monica's Clover Field under the watchful eye of Charlie Lalotte, a 22-year-old Army-trained aviator fresh out of the National Guard. Noah Dietrich claimed that Hughes had received some previous instruction from J.B. Alexander in 1925, but he apparently kept this secret from Lalotte in order to get as much dual as possible before he soloed.

Lalotte was to go on record: *'He wasn't interested in just soloing, he was there to learn all he could from me. He was a terrific pilot, not just a dumb cluck.'*

By December 1927 Hughes had gained his private pilot's licence. By the end of August 1928 he had his commercial rating, to which later he would add multi engine and instrument ratings.

Hughes was to put his flying skills to good use during the making of *Hell's Angels* when it was said that Hughes would not ask his pilots to do anything he himself could not do. One day Frank Clarke, one of the film's stunt pilots, balked when Hughes asked him to make a steep left turn after take-off in a Thomas Morse Scout and dive past the cameras. *'Howard, the gyroscopic forces of a LeRhone rotary engine will force you into the ground in such a turn'*. Despite this warning, Hughes climbed in, started the engines, adjusted his goggles, took off, turned left - and promptly crashed. He recovered, but the accident left its mark: a crushed cheekbone would trouble him in later years.

Hughes' insatiable desire to learn all he could about flying led to him taking a job in 1932 with American Airways working for three months under the name of Charles Howard, flying as a co-pilot on the Fort Worth to Cleveland run flying an old Ford trimotor. His interest in flying was growing rapidly and he began to re-design an Army Boeing 12 for racing. To help him take it apart and improve it he hired Glen Odekirk, a pilot of his own age who was weathering the depression as a humble mechanic in the Pacific Airmotive shop at Burbank on the north side of Los Angeles. Odekirk learned to adjust to Hughes, and late in 1932 found himself starting a long process of modifications to a one-off called the Boeing 100A. This was basically a member of the prolific P-12/F4B family, with Pratt & Whitney R-1540 Wasp engine and initially

Howard Hughes unloads passenger baggage in 1932 while working as a co-pilot for American Airways supposedly under the name of 'Charles Howard'. This is somewhat hard to believe, as his Transport Pilot's Licence No. 4223, issued on 15 May 1929 was clearly made out in the name of 'Howard R Hughes' and would have certainly been something that American Airways would have checked!

capable of about 185 m.p.h. Registered N247K, it was possibly the fastest private vehicle in the world. Then Hughes and Odekirk got to work on it. It had been specially made for Hughes with two seats, but was soon modified into a single-seat racer, with every conceivable aerodynamic improvement from a new spatted landing gear to a long-chord cowling over a highly souped-up engine.

The Boeing was tuned to give as much power as Odekirk could wring from it - perhaps 700 hp - certainly enough so that Hughes was able to race away with the Sportsman's Cup at a speed more than one-third higher than the maximum possible by the basically similar Boeings of the Army or Navy. For good measure Hughes treated the crowd to a remarkably exhaustive and polished display of aerobatics before landing.

In those days a special aura of glamour and adventure attached to fliers, and this, combined with Hughes's tall, dark, youthful good looks, wealth, and reputation as a moviemaker, made him much sought after as a guest on the estates of the wealthy. He was a frequent guest of Sherman Fairchild, son of the founder, first president, and first chairman of the board of International Business Machines - IBM. Fairchild was handsome, single, wealthy and, just like Hughes, fascinated by the technology of aviation. Fairchild was already a leading figure in the aviation industry. In 1920 he had successfully demonstrated his automatic aerial camera and formed a company to manufacture it. Worldwide acceptance of the camera gave rise to a wholly new field of aerial surveying and mapping, which in turn established the need for an aircraft that could provide a suitable platform for aerial photography. After five years of attempting to modify existing aircraft to meet the specific requirements of a good photoplane, Fairchild characteristically decided to build his own.

By 1928 Fairchild's company was the nation's largest builder of cabin monoplanes. One of his aircraft, FC2-W *City of New York*, held the record for the fastest trip around the world while another FC2-W, *The Stars and Stripes*, made the first flight from the Antarctic continent while serving with the 1928 Byrd expedition. Still another of the series successfully explored the unmapped interior of New Guinea, a most unlikely environment for an aircraft in 1928. For its time the Fairchild cabin monoplane was a superb

Howard Hughes
standing alongside his
much-modified Boeing
12/100

bush country aircraft.

Hughes and Fairchild, who had first met at a party in New York a couple of years before, had much in common, and their long friendship was to span two decades. In the years that followed, Hughes often flew to visit Fairchild at his Long Island estate. One of Fairchild's projects would play an important part in the Hercules story.

As the summer of 1933 drew to a close, Hughes asked Odekirk to come down to the Drake. A party was going on with Sherman Fairchild and other friends of Hughes when Odekirk arrived. Midway through the party, Hughes said, *'Odie, I want to talk to you.'* In a back bedroom Hughes told him he was leaving for Europe the next day, would be gone a couple of months, and wanted Odekirk to have some work done on the aircraft while he was gone.

While in Europe Hughes bought the 286-foot British-built twin screw steam yacht *Rover,* which he registered in Panama, renamed *Southern Cross*, and had delivered to him in Florida.

When the *Southern Cross* tied up in Miami, Hughes tried living aboard, but could not adapt to the maritime life, so in two or three days he had moved back to his hotel. As usual, all the finance came from Hughes Tool Company. It was the middle of the depression, and even Toolco was not immune to hard times, but Hughes' idea of being frugal was different to everyone else. Odekirk is supposed to have overhead a conversation between Hughes and Noah Dietrch in Houston when they were sharing a suite in Palm Beach. *'Well, Noah, I know things are a little rough... I'm trying to be careful of what I'm spending.... In fact, I'm trying to save some money. I've just moved from a thirty-dollar-a-day room to a twenty!* At the time Hughes had just bought the world's fifth largest private yacht, was paying a crew of thirty to man it, plus was maintaining two aircraft in Florida!

Glenn 'Odie' Odekirk (*b.* 9 May 1905 - *d.*12 January 1987)

Hughes's friendship with Sherman Fairchild sheds light on both his personality and on his inclinations as an aviation pioneer searching for success. The two men were good friends and in frequent contact with each other from the early 1930s to the 1950s when Hughes began to retreat from public view. Their personalities were similar - both were wealthy, attractive, single, and shared an interest in beautiful women. And both were fascinated by, involved with, and had high aptitude for technology, particularly that associated with aviation.

Sherman Mills Fairchild (*b.* 7 April 1896 – *d.* 28 March 1971)

Perhaps the most important trait they shared was a waywardness in their powers of concentration. It seems that both men were capable of intense concentration on a single interest: but neither could retain this interest for long. Through the 1930s Hughes turned from moviemaking to aviation to moviemaking again, while Fairchild shifted from aviation to photography to boats to the tennis courts and back to aviation. As a result, the aircraft industry did not take either too seriously.

Aviation professionals such as Donald Douglas and 'Hap' Arnold, whole lives were given over to aviation saw Fairchild and Hughes as amateurs - gifted most definitely, but more interested in pursuing new ideas than in concentrating on the central business of the industry: building more and better aircraft than the next fellow.

The steam-yacht *Rover,* built by Alexander Stephen & Sons of Glasgow, Scotland. The vessel was the largest steam yacht ever built on the Clyde. She was 265 feet long with a beam of 40 feet. From the time of her delivery in July 1930 Lord Inchcape spent much of his time aboard her and at his estate at Glenapp Castle. *Rover* was schooner rigged with clipper bow and counter stern making her a most handsome vessel. Lord Inchcape died aboard the *Rover* at Monte Carlo in 1932 and the vessel was sold Howard Hughes and re-named *'Southern Cross'*

It was typical of Sherman Fairchild that he should become much more interested in an idea such as Duramold, a product that he pursued to the detriment of the Fairchild company as a whole than in developing a product more in line with the company's experience - and therefore more likely to make money for it. Similarly, none of the major manufacturers would have spent nearly two years perfecting the Hughes Special, a machine that had been built for a single goal – speed - and which could therefore never be sold in quantity. There was a feeling within the industry that Hughes and Fairchild - especially Hughes - were simply dilettantes who could afford to play around with aircraft ideas.

Certainly the aviation industry admired their contributions. With the H-1 Hughes broke the world's landplane speed record and he added significantly to aeronautical know-how in the process. With Duramold, Fairchild pioneered such techniques as electronic gluing, and as a structural material Duramold presaged modern sandwich and honeycomb structures. Nevertheless, the activities of neither man led directly to a major business success; for neither had yet found the proper mix of technology and market readiness necessary at any given time for the economic success of an aircraft design.

Aviation success then, as now, is an economic matter; an aircraft had to be marketable, which meant that it had to be useful to a significant number of aircraft buyers. By the outbreak of World War Two it was apparent that such success would depend almost exclusively on building aircraft useful in war. Both Fairchild and Hughes wanted to succeed in the conventional business fashion. Hughes made persistent attempts to sell aircraft designs to the skeptical professionals at Wright Field. But by concentrating to the point of obsession on the technology of speed, on breaking records, and on doing things his own way, Hughes blew his chances.

The two men's careers diverged where their personalities did. Fairchild was a reasonable man. If someone gave him a good reason why an idea would not work, he would agree and move on to another idea. He seemed to have been better able to learn from his mistakes than Hughes and was

Howard Hughes as a very young and very famous celebrity. He is seen here with record-breaking American aviator Roscoe Turner (in flying helmet) and other aviation enthusiasts.

more willing to admit that he might be wrong. Howard Hughes was different - once Hughes had an idea in mind, nothing could change it. Unlike Fairchild, he did not know his own limitations. He was like a cannon ball: once fired, he blasted all obstacles from his course.

Why don't you build one yourself?

Despite all the work that was done on the Boeing, Hughes was still not satisfied and another Hughes legend was born.

Allegedly Odekirk had become exasperated with Hughes' constant wish to improve the Boeing that he had said, not really thinking of the consequences, *'The only way you'll ever get a ship to please you 100 per cent will be to design it yourself'*. In 1934 Hughes' spirit was constantly airborne and he very soon decided that Odekirk's sarcastic comment was the key to what he would be doing in the next year or two. He calmly told Odekirk to set up a workshop where the first Hughes aircraft could be created.

From the outset his aim was to build an aeroplane to go as fast as possible. He could use any engine he liked, and if he wished could spend much more than a mere million dollars. In March 1934 he began to set up an organisation in Charles Babb's hangar at Grand Central Air Terminal, which was then the LA airport at Glendale, where Odekirk was already laying in a foundation of tools and raw material.

He called on some former classmates at the California Institute of Technology Pasadena, California for assistance. Next, he set up Hughes Aircraft Company and hired Dick Palmer to head the project under his close supervision. By mid-1934 the total team of eighteen were working full-time in a walled-off section of the hangar. Before they joined Hughes they had agreed to preserve complete secrecy. Hughes, worked out preliminary designs and had a model built and tested in the wind tunnel at California Institute of Technology.

The tests proved it was a remarkable design and construction began. The aircraft included such features as retractable landing gear and an all-metal fuselage put together with flush rivets to reduce drag. There were other unique features such as the oversized constant speed propeller and fully cowled engines that, when put together, added up to a most ingenious concept. The completed machine was designated the H-1 for the Hughes

Aircraft Company's first aeroplane. Hughes was ready to make an assault on the world land plane speed record. At the time he called on his friends, Amelia Earhart and Paul Mantz, to act as observers. In 1935, Hughes went to the deserted Irvine Ranch near Santa Ana, California to make his attempt to return the record to the United States. With Joseph Nikrent as the official timer, Hughes began his runs to break Frenchman Delmotte's record of 314.3 miles per hour.

Hughes made his first run over the measured mile course and came up short-flying at 302 miles per hour. His second run was outstanding, reaching a speed of 350 miles per hour, but it was disqualified because the timer said he was in a shallow dive during the actual run. His third run that day was officially clocked at 334 miles per hour - easily fast enough to break the record.

On the next day, as soon as everyone else was ready, Hughes took off. After ten minutes he had made four runs, and then added a fifth, because one run had been a slow 337 m.p.h. The judges radioed that the mean speed of the four good runs was 352.39 mph. This was a bigger advance than the

Above: Howard Hughes sitting on the roped off wreckage of the downed H-1 shortly after the accident. Some reports in the press said it was a blocked fuel line that caused it - others admitted that he simply ran out of fuel.

Right: the re-built, modified version of the H-1 with longer wings and enclosed cockpit for high-altitude flying.

landplane speed record had ever before seen, but Hughes was confident he could do 10 m.p.h. better. He came in for a sixth run, undoubtedly the fastest yet. Just as he entered the course the engine cut dead. To those on the ground the sudden eerie sound of a 365 m.p.h. glider was horrifying.

Hughes was desperately switching fuel tanks. He and Palmer had somehow found room for 250 US gallons, but this was of academic interest if the supply pipe was blocked. Hughes shut off the fuel and ignition, lowered the gear and decided that it would be suicidal to try to turn back to the field. He looked around, and decided the best bet was a large ploughed beetfield, made a further wise decision in retracting the gear again and finally brought off a well-judged belly landing-almost certainly the fastest the world had then seen. When the dust cleared, the apprehensive crowd racing to the wreckage found Hughes sitting on the fuselage writing notes.

Hughes ordered the H-1 rebuilt with larger wings for long-distance flying. He had big plans for it - plans that would bring him great honours and contribute enormously to the advancement of aviation. In the meantime he 'made use of' a Northrop Gamma from famed aviatrix Jacqueline Cochran who had recently purchased the Gamma, a sleek advanced monoplane she was readying for the Bendix race. Hughes calculated that if he replaced the 1535 engine Cochran had on the machine with the latest Wright Cyclone R-1820G 850 horsepower engine coupled with a Hamilton Standard variable pitch propeller, he could easily better Roscoe Turner's existing record.

He offered to rent the Gamma from her for nearly as much as she had paid for it. Meanwhile, Hughes made eleven flights as a Douglas DC-2 copilot on TWA's transcontinental runs during 1935, apparently to build his transcontinental experience in preparation for the record attempt. Finally,

Howard Hughes climbs out of the Northrop Gamma at Newark, New Jersey.

on 13 January 1936 Hughes flew the modified Gamma from Burbank to Newark in nine hours and twenty-seven minutes at an average speed of 259.1 mph for a new record. He set a new record flying from New York to Miami by making the flight in 4 hours 21 minutes. He then set another record flying from Chicago to Los Angeles, in the face of heavy head winds, in the record time of 8 hours and 10 minutes.

The *New York Times* said of Hughes, *'When an amateur pilot in the space of a few months can break the world land plane speed record, and lop half an hour off the transcontinental record, he is a pilot to be reckoned with.'*

For his achievements Hughes was awarded the Harmon trophy. On 20 January 1937 enroute to the presentation ceremony he flew a revamped H-1, now fitted with a longer wing and a new Pratt and Whitney R-1535 Wasp engine of 700 hp, from Burbank to Newark in seven hours, twenty-eight minutes and thirty-five seconds. Hughes built two sets of wings for the H-1, one with a span of only twenty-five feet that he used to set the closed course record, and the other with a span of thirty-one feet nine inches that he used for his long-distance runs. The wings were of wood and the fuselage was aluminum. The racer averaged 327.15 miles per hour over the 2,490-mile course for a record that was to stand for ten years.

It was here that another minor mystery appears - minor as to the event, but major as to it's later implications. People close to Hughes have gone on record stating that while he was still on the East Coast after his record-breaking transcontinental flight in the H-1 he was telephoned by General Oliver P. Echols, Commander of the Army Air Corps' Wright Field in Dayton, Ohio, the centre for Air Corps testing and procurement. Echols told Hughes that the Air Corps were interested in the H-1 because it was faster than anything they had at the time. *'Can you stop by and let us see it on your way back to California?'* Echols is supposed to have asked Hughes, who agreed and Echols arranged for a group of top brass to be on hand to meet him.

Colonel (later General) Oliver Patton Echols (*b.* 4 March 1892 – *d.* 15 May 1954). He and Howard Hughes would clash a number of times.

According to Noah Dietrich, there now occurred the first of several incidents that would poison the minds of key Army Air Corps officers against Hughes. He overflew Wright Field, re-fuelled in Chicago, and continued on to California. Echols, who later became Chief of Air Corps procurement, never forgot the snub, vowing that Hughes would never get a 'dime's worth of business' from him. Hughes is supposed to have told Dietrich that he just forgot to stop in Dayton. Dietrich thought the snub was intentional, that he simply *'didn't want those generals snooping around his airplane and stealing his ideas.'*

Such an incident did occur, according to the testimony given in the 1947 Senate hearings, but not in the way Dietrich recalls in his book. According to information in Hughes's logbooks Hughes did not fly the racer home. The aircraft sat in Newark until Allen Russell, corporate pilot for William Randolph Hearst, flew it back to Burbank.

Hughes had become one of aviation's brightest, ascending stars - as both as an innovator and an engineer as well as a pilot. His flights had not been publicity stunts, they were scientific studies intended to advance the knowledge of powered flight. Everything he had accomplished up to now was leading to his next endeavour, the pinnacle of his career as an aviator - a flight around the world.

The record for a flight around the world belonged to his friend, Wiley Post, who made the flight in 7 days, 18 hours and 49.5 minutes in a Lockheed Vega. The flight Hughes intended was not to be just for the record, but it would be the most important scientific flight every made. He wanted to use the latest in radio communication, weather forecasting and reporting, and navigational equipment.

At first he planned to use an S-43 Sikorsky amphibian that he bought in 1937. But because of the slow speed he shopped around for something more suitable and selected a twin-engine, low-wing Lockheed Model 14.

The attempt would be launched from the greatest world stage that year - the New York World's Fair in 1938. Preparations for the flight took two years. He was meticulous about every detail and tried to provide for every possible contingency. Because the Lodestar was a landplane he had eighty pounds of Ping-Pong balls stuffed into all the empty recesses of the wings and fuselage so the aircraft would float if it were forced down at sea. He had Odekirk and his hangar crew strip the aircraft of all non-essential weight and install extra fuel tanks, all of which were lined with neoprene to make them self-sealing. Hughes also carried ethyl on the flight to mix with low-test gas if he were unable to obtain the proper octane fuel along the way. Hughes spared no expense in equipping the aircraft with the latest blind flying, communication, and navigation equipment, including the newly-perfected Sperry Gyro Pilot, which could automatically maintain level flight and help to hold accurate headings on the long over-water legs of the flight.

Hughes, through William C. Rockefeller, would make use of a

With four crewmen, Howard Hughes established a new record for around-the-world flight of three days, nineteen hours and seventeen minutes on 14 July 1938, thus shattering Wiley Post's solo record of seven days, eighteen hours and forty-nine minutes which was established in 1933..

Add New Chapter to Aviation History With Record Flight

The plane

Richard Stoddart   Harry P. Connor   Howard Hughes   Lieut. T. L. Thurlow   Edward Lund

worldwide weather forecasting and reporting system - reports from around the world would be funneled into the New York World's Fair and re-transmitted by radio relay to Hughes wherever he was.

Rockefeller established a 24-hour weather forecast centre at the New York World's Fair grounds with the help of its manager, Grover Whalen. The latest teletypes funneled data from London, Paris, Berlin, Rome, Moscow, Manila, Honolulu, San Francisco, and Washington. Daily synoptic charts provided forecasts of wind direction and strength, temperatures, ceilings, visibilities, and storm warnings. Hughes arranged for certain radio stations throughout the world to broadcast, periodically, signals that his navigator/radioman could employ in pinpointing their location at anytime, anyplace in the world. The aircraft was equipped with the latest instrumentation, Nav-Com. equipment and a 15-watt transmitter. He used his own initials for call letters. He also set aside 30 channels on the radio for every conceivable emergency.

Hughes had already demonstrated a talent of chosing the right people for the job - so it was with the crew of the Lodestar. Radioman Richard Stoddard, 38, was a former shipboard radio operator, also a licensed aircraft pilot, and an NBC communications engineer. Navigator Thomas Thurlow, 33, was an Army pilot and navigator noted for his development of special navigation instruments. The combination copilot-alternate navigator was Harry P. McLean Conner, an expert navigator with trans-ocean experience on previous record-breaking flights.

The flight engineer was to have been Glen Odekirk. But he had been working long hours with little sleep to get the machine ready for the flight; he had lost thirty-five pounds and was worn out. Nevertheless, he told Hughes that if weather delayed the departure long enough for him to get some rest, he would go. But in New York Hughes decided he wanted all new cylinders on the engines, so Odekirk had to work day and night with Wright Aeronautical people to get the job done. Consequently he was replaced by Ed Lund, a Hughes employee who was an expert on aircraft engine maintenance.

Because the primary aim of the flight was to publicise the reliability and safety of flying, Hughes astutely arranged for it to represent the New York World's Fair, scheduled to open in 1939. He could thus avail himself of the services of Grover Whalen, president of the fair and New York's official greeter. Whalen cranked up a high-powered campaign that included radio, movie and press coverage of the official departure and arrival ceremonies, and flight progress releases. Millions of people around the world followed the flight avidly.

The aeroplane and crew were ready. Sixteen hours after the take-off from Floyd Bennett Field, the Lodestar landed in Le Bourget, France. Virtually every stop along the way was record time for a flight from the previous stop. Tail winds assisted them on the way to Moscow. The stops after Moscow were Omsk, Yakutsk and then back onto American soil again at Fairbanks, Alaska, and then home.

When Hughes landed in New York on 13 July 1938, ninety-one hours after his departure, a surging crowd made it difficult for Mayor Fiorello La Guardia and Grover Whalen to officially welcome him back. La Guardia and Whalen accompanied him on the traditional ticker tape parade up Broadway where, according to the New York sanitation department, the

Two views of the arrival back of Hughes' Lockheed 14 Lodestar at New York's Floyd Bennet Field following the record-breaking around the world flight Howard Hughes and the crew were met by well over 30,000 spectators and 1,000 police officers to control the throng!

populace threw a greater weight of confetti and ticker tape in his honour than had been thrown during Lindbergh's reception after his solo transatlantic flight to Paris.

The honours began to pour in and Howard Hughes was about to take his place in the history of aviation. First, the National Aeronautics Association named him Aviator of the Year. Then the Congress voted him a special congressional medal for his achievements as a pilot. In December 1939, the name of Howard Hughes was added to the list of aviation's immortals when he was awarded the Collier Trophy for the greatest achievement in aviation in America. This most coveted award in aviation was presented to him personally by the President of the United States, Franklin D. Roosevelt.

Hughes next planned to fly around the world on a goodwill tour of major world capitals in the first pressurised, high-altitude transport, the Boeing Model 307 Stratoliner. In the process, he bailed Boeing out of a difficult situation. Although ten Stratoliners had been ordered - four for Pan American and six for TWA - and were in various stages of construction, Boeing had cash flow problems. It had spent a lot of money developing the Model 299 prototype of the famous B-17 Flying Fortress as its entry in an Air Corps bomber competition in the mid-1935.

As part of its effort to recoup, Boeing went after airline money with its Stratoliner design, which incorporated the wings, powerplants, nacelles, and tail surfaces of the 299. But by the time Hughes and Odekirk went to Seattle to arrange payment for the TWA purchase, Boeing was hurting. Hughes now owned enough TWA stock to have controlling interest, and part of the deal was that Hughes was to get one of the six aircraft that

TWA had ordered. Hughes arranged that the Hughes Tool Company would make weekly partial payments on TWA's six aircraft of around $100,000 a week.

Germany's attack on Poland 1 September 1939 began World War Two and ended Hughes's plans for another world flight. At first German air power ruled the skies as a key element of the blitzkrieg that carried the Nazi war machine to success after success. Shaken, the United States Army Air Corps (USAAC) launched a crash programme to develop aircraft superior to that of Germany, even though they did not enter the war until December 1941. It was then that Hughes decided to enter the competition for the development of a high-performance interceptor.

Designated as the D-2 project it envisaged a twin-boom, twin-engined aircraft with a relatively small central crew nacelle. No effort was spared to minimise aerodynamic drag. Most of the airframe of the D-2 was to be made of Duramold plywood. This material was advantageous from an aerodynamic and a metals-shortage standpoint, but was difficult to work.

The war in Europe caused all thoughts of a record-breaking round-the-world flight to be abandoned. Aircraft engines and specialised aviation equipment urgently needed for rearmament were not made available for the purpose of setting records. In order to keep his project going, in December 1939 Hughes offered to sell the drawings and data for the D-2 to the USAAC in the hopes of attracting a military contract. He hinted that the design might be made into a 'pursuit-type airplane', although at that time the D-2 had no military role envisaged.

Hughes seems to have succeeded in interesting the Army in his project, since in 1940 the USAAC informed him that there was no objection to his

Howard Hughes inspects one of the engines of the D-2 prototype. Much of the structure was made from Duramold - the process developed by Sherman Fairchild's Fairchild Corporation.

purchase of a pair of Wright Tornado engines. However, the USAAF decided later to divert these engines to Lockheed for its XP-58 Chain Lightning project, leaving Hughes without engines for his D-2. Hughes was forced to switch to a pair of Pratt & Whitney R-2800-49s for the D-2, and these engines arrived at Hughes in March of 1942.

Initially, the aircraft was to have been a taildragger, but the landing gear was later changed to a tricycle configuration. The main undercarriage units retracted rearwards into the twin booms and the nosewheel retraced rearwards and rotated 90 degrees to lie flat in the small central fuselage.

The design was somewhat similar to the Lockheed P-38 Lightning that won the 1939 USAAC design competition. Hughes later testified to the U.S.

Left to right: Stanley Bell, Kenneth Ridley, Howard Hughes, Odie Odekirk. The date is thought to be July 1943. *(Simon Peters Collection).*

There are very few views of the D-2 outside. As can be seen, there are certain similarities with the Lockheed P-38 Lightning, which Hughes later claimed they 'pirated' from him. *(Simon Peters Collection).*

Senate that Lockheed had stolen his design, although this was refuted by many others. Rather than abandon the project, as he later recounted in the 1947 Senate investigation, he *'decided to design and build from the ground up, and with my own money, an entirely new airplane which would be so sensational in its performance that the Army would have to accept it.'*

Memos between Hughes and the AAC seem to indicate that he was changing his mind about the mission for the D-2. In May 1940, he was no longer referring to a 'pursuit-type aircraft', but a 'Duramold bombardment aircraft'. The mission envisaged for the D-2 seems to have changed yet again in May 1941, the aircraft now being pictured as a bomber escort. To further muddy the waters, in June 1942, the USAAF seems to have referred to the aircraft as the 'P-73' in one communique as a fighter version of the D-2, and as the 'XA-37' for the attack version in another, indicating that there was really no clearly defined military role for the aircraft. In reality, the D-2 could not carry enough bombs internally to make it a useful attack aircraft and was not sufficiently manoeuvrable to make it a useful fighter. These designations may have been little more than place holders.

Hughes suspected - and not without some justification - that one reason he had not been awarded the interceptor contract was that he lacked the necessary production capacity. Now with the D-2 under development they had to find space for a plant.

Odekirk was ordered to overfly the Los Angeles area to see if he could discover any vacant real estate. He narrowed it down to two locations, then

he and Hughes flew over the likely properties together.

First was an area in the San Fernando Valley that had potential multi-directional runways on a two-square-mile piece of land, but Hughes thought it was too small. They flew on, approaching the Pacific Ocean shoreline west of Los Angeles. Odekirk pointed at a long strip of land running perpendicular to the coast, a huge area - some said as much as 1,000 acres - of the Ballona wetlands south of Jefferson Avenue in south-west Culver City. Here was room for a 9,500-foot runway. Hughes recognised the area as one of the few large tracts of remaining undeveloped land in Los Angeles, but the high water table made it necessary to sink fifty foot pilings into the wetlands to support Hughes' buildings and reroute the course of the Centinela River, which flowed

4 July 1944, and Howard Hughes talks over details of the D-2.

The sleek lines of the D-2 are on display in this illicitly gained picture. *(Simon Peters Collection).*

through the site every spring and flooded it.'Buy it' said Hughes. Allegedly he paid half a million dollars for the site.

Plant construction started early in 1941 with a 60,000 square foot air conditioned, humidity-controlled aircraft plant with an adjacent grass runway. But what with building the D-2, buying the land - Hughes wanted everything that was vacant to be bought to keep people from getting too near the airport - and building the first three buildings of the future Hughes Aircraft plant, they were spending, not making, lots of money - and, as usual, it was Hughes Tool Company that was paying the bills. Noah Dietrich in Houston didn't think that was any way to run a business.

By Pearl Harbor Hughes had his new plant and was ready for bigger things. But his unorthodox business methods, obsessive secrecy, and desire to do things his own way contributed to his continuing failure to sell his D-2, or any other machine, to the Air Corps.

The D-2 was built in secret with Glenn Odekirk, providing engineering inputs. The secrecy further alienated USAAC officers, especially when Hughes denied Materiel Command access to the plant. The USAAC had requested information about the project's progress, but did not enter into a formal contract until 1944. Final assembly and flight testing occurred at the Hughes Harper Dry Lake facility in the Mojave Desert.

Then along came Henry J. Kaiser in August 1942, and the story of the Spruce Goose began.

Work started on the new building complex at Hughes Aircraft Culver City Plant in the spring of 1943, watched over by barrage balloons scattered around the area.

Partners

It is difficult to imagine two more different men than the tall, slender, introverted Howard Hughes and the short, roly-poly, extroverted Henry Kaiser. Hughes shunned any publicity apart from what could be manufactured, controlled and manipulated on his own terms; Kaiser sought it at any count, no matter what the terms and conditions. Hughes was a perfectionist; Kaiser was a pragmatist. Kaiser was hard-driving, punctual, and lived according to a rigorous schedule. Hughes was dilatory, kept no regular office hours and seemed to run his businesses by telephone from wherever he might be at the moment.

When Kaiser wanted to contact Hughes, his staff would have to call Hughes's Romaine Street office in Hollywood and call again, and then again. Each time the young minions on duty at Hughes would be vague about the whereabouts of their master and would politely ask the caller to leave a message. Hughes could never be reached on the spur of the moment. It's alleged that his uncle Rupert once complained that *'I can get through to the Almighty by dropping to my knees, but I don't know how to get in touch with Howard.'*

A Partnership Is Formed.

Glenn Odekirk was listening to the radio news on Sunday 9 August which was talking of Kaiser's meetings with members of Congress to promote his idea for giant flying boats. Sensing an opportunity, Odekirk immediately dialed Hughes's unlisted number in Bel Air, California. Odekirk discussed it with Hughes, the result being that they would consider designing it themselves and let Kaiser's team build it.

A week later Odekirk put a call through to Kaiser on the East Coast. Kaiser went for the idea in a big way. His voice was excited. He told Odekirk he was leaving for Oakland by train that night, and would arrive there Wednesday afternoon. During the train jouney Kaiser supposedly called Odekirk from stops along the way. By the time the train reached Oakland they were on a first-name basis, inviting the pair of them up for the launch of their first Liberty boat built in just ten days. *'We'll be on national news and we'll announce we're going to join forces to build the biggest flying boats ever seen.'* Odekirk was taken aback by the abrupt proposal.

Hughes had been in San Francisco all week for a pre-showing of his latest motion picture *The Outlaw.* starring Jane Russell. He was suffering from a bad cold that had grown worse and now was almost the flu. He was holed up in the Fairmont Hotel under an assumed name.

As soon as Kaiser learned of Hughes's whereabouts from Odekirk he hung up, called for his car and was rushed across the San Francisco Bay Bridge from his Oakland Headquarters to the Fairmont Hotel, where his peppery enthusiasm was pitted against Hughes's low spirits.

Kaiser was twenty-three years older than Hughes, and went straight into an inspirational sales pitch. Even the skeptical Hughes, who already knew something of the problems and time requirements involved, was intrigued by the possibility of designing the world's largest aircraft.

After several days of talks in which Odekirk and Kaiser's son Edgar participated, they reached a handshake agreement. Hughes would design a flying boat; Kaiser would build it.

On Sunday evening Kaiser, elated, announced the agreement to the press. On Monday morning Hughes viewed the headlines with distaste. The press release bound him publicly to goals and completion dates dreamed up by Kaiser and not studied by Hughes. Kaiser said that they had agreed to go forward in fulfilling *'...the most ambitious aviation program the world has ever known.'*

Hughes was not the only one who was worried. Officials in both the War and Navy Departments also concerned that a go-ahead might be given to such a project during the most critical period of the war. This did not necessarily mean that Kaiser was wrong - after all, the Navy had opposed at first the Kaiser proposal to build a fleet of escort carriers on merchant ship hulls - and that programme was a great success.

Chairman of the War Production Board Donald Nelson sent Grover Loening to the San Francisco Bay area to inspect Kaiser's facilities on 22 and 23 August which overlapped Kaiser's meetings with Hughes. Kaiser took the opportunity to ask Loening what he thought of Howard Hughes. According to Kaiser, Loening said that he thought Hughes was a brilliant engineer and that his aircraft work *'...was just about the best that we had in the factories of America'*.

Loening thought that the machinery and general tools of Kaiser's Bay Area yards were not suitable for aircraft construction. He was also concerned that no serious aircraft design work had been done by any Kaiser engineer, and that none of its engineers had been sent for aeronautical training, even though California universities had offered

Concept No.1 - the first sketch of the HK-1, showing a single-hull, twin boom design, reminiscent of the D-2, but with eight engines.

their facilities to the Kaiser organisation.

In his report to Nelson, Loening noted that during his meeting in Los Angeles on Monday 24 August with the committee of aircraft builders, Kaiser had said that he greatly needed the assistance of the aircraft industry in designing a large, possibly 200-ton, aircraft for him. When it was pointed out to Kaiser that this was physically impossible due to other demands on the aircraft industry, Kaiser immediately took this as a personal affront and said that he intended to enter into partnership with Howard Hughes in order to use the Hughes engineering group for design work. Kaiser then denounced the committee, saying *'The whole aeronautical industry is ganging up on me and trying to prevent me from getting into the business.'*

Loening concluded from his West Coast visit that Kaiser had planned to make the aircraft industry design his aircraft and for the government build his factory so that *'...Mr. Kaiser's own contribution to adding aircraft facilities at the present crucial time would have been exactly nothing but newspaper buildup'*. Nevertheless, it was clear that Loening was impressed with the efficient organisation of Kaiser's enterprises, their orderly operations, the boldness of the assembly line methods, and the obvious competence of the men who worked for Kaiser. In his report to Nelson he further recommended that *'...since Mr. Kaiser has become associated with Mr. Hughes and has in this way acquired some aeronautical staff, although not nearly large enough, that he be requested to design a new cargo carrying seaplane of at least one hundred tons gross weight size. An order should be given Hughes for the design only, in order to determine the fitness of himself and his organization to do work of this character'*.

He also recommended to Nelson that *'...during the time that this design is being prepared, Mr. Kaiser be required to present to the government detailed plans of a factory that would use no strategic materials and of the assembly of a staff of engineers and factory workmen that would be trained from his own resources and not taken from the existing aircraft industry'*.

Kaiser gained powerful support from Jesse Jones, who as Secretary of Commerce had built a financial empire in Texas. He wielded wide influence because of his control of both Commerce and the Federal Loan Agency and through his ties with southerners on Capitol Hill. Some of Nelson's advisers, including Merrill Meigs, tried to dissuade Nelson from backing Kaiser. But others pushed him to decide in favour of building the giant flying boats. Col. Roy B. Lord, a strong advocate of transport aviation, wrote to Nelson as Chairman of the War Production Board on 31 August expressing regret that Kaiser's proposal had run into so many snags and delays: *'...It is apparent that nothing constructive can be accomplished by continuing the committee of top men from the aircraft industry since they are naturally against permitting a new competitor to manufacture an air cargo plane which may put this competitor in the lead during the post-war period.'*

'With the strong tie-up between Kaiser Co., Inc. and Howard Hughes who has one of the best reputations as a designer in the industry, and with the general public as favorable to Kaiser as is indicated by the public polls and in the newspapers, I feel that it would be a grave mistake

not to at least let Kaiser design and construct several of the large planes'.

Colonel Lord's letter was accompanied by a covering memorandum strongly endorsing Lord's recommendations and providing additional arguments designed to sway Nelson in Kaiser's favour. The memorandum's author, Nelson's assistant, Edwin A. Locke, Jnr stated:

1. *'If you take the advice of the aircraft industry, you will tell Kaiser to forget about producing airplanes and concentrate on providing raw materials. In that event, every newspaper in the country will 'go to town' in favor of Kaiser and against the WPB.*

2. *'I think the newspapers will be right. The general public and many of us in the war agencies have been deeply impressed by the Kaiser*

Above: Concept No.2 - twin hulls, twin booms and seven engines.

Left: Concept No.3 - same basic design, but with six engines

proposal and believe that under present critical conditions this country cannot afford to deny it, especially when presented by a man with a record such as Kaiser's for accomplishing the impossible.'

3. 'If you follow Colonel Lord's recommendation, what do you stand to lose: If the planes fail, the loss will be perhaps two or three million dollars and an infinitely small amount of materiel. If Kaiser succeeds, the benefits to the war effort will be great.'

4. 'A contract to Kaiser will provide some very salutary competition for the Army, the Navy and the aircraft manufacturers'.

In fairness to the established aircraft manufacturers, it should be noted that they had abandoned many peacetime competitive practices in order to produce Army and Navy aircraft in the shortest possible time. They cooperated with the automotive industry in its aircraft activities and worked together through Aircraft War Production Councils. But it was natural they thought their own production more important to the war effort than the Kaiser scheme; such a conclusion was warranted by the facts as they saw them.

Nevertheless, Nelson was convinced he would be remiss if he did not give the flying boat project a go-ahead, and the press and public would have been quick to agree. Moreover, Kaiser was applying pressure through his influential and powerful friends in Washington, a process later described as '*...putting the heat of hell on Washington'*.

As a final step before issuing a letter of intent, Nelson had Harold Talbott check with the War and Navy Departments to make sure that the proposed project would not interfere with current programmes. On 12 September 1942, Towers, responding for the Navy, and Robert Lovett, Assistant Secretary of War for Air, responding for the Army, signed letters to Nelson agreeing to make standard engines and instruments available for three cargo planes as proposed. Both men stressed that the Kaiser-Hughes organisation should be specifically prevented from hiring away key personnel from existing governmental aircraft contractors. This restriction - along with the denial of the use of strategically important

Concept No.4 - single hull and tail, and six engines. The aircraft is side-loading.

Concept No.5 - a
refined version of
Concept No.4.

materials such as aluminium caused many problems.

All Kaiser and Hughes had to do was to build an aeroplane bigger than any the world had ever known and do it without aluminium and critical materials and without hiring any aeronautical engineers or specially skilled personnel from other companies. As a result, Hughes decided it would be best to train their own engineers anyway.

As soon as the letter of intent had been issued, Kaiser took a team of production men and engineers to the Hughes Aircraft plant in Culver City. Odekirk had only a small office in front of their new factory, as their office building had not yet been built. He was overwhelmed by the size of Kaiser's groups who were planning to build a waterway from the plant out into the ocean to launch the flying boats and with the speed of which they wanted to work. After suffering weeks of frustration, the Kaiser engineers and production people returned to Kaiser headquarters in Oakland. But the two 'partners' did manage to set up the Kaiser-Hughes Corporation with offices in Culver City.

Despite the joint name, Hughes continued to go it alone. He began making frequent flights to Lake Mead, Nevada to experiment with takeoffs and landings in the Sikorsky flying boat he had purchased in 1937. Hughes's reputation grew for immersing himself in detail, for tinkering with designs and specifications, for being difficult to locate when needed, and for failing to follow accepted business practices. Key people in government and the military began to feel that the war could not wait for a dilatory Howard Hughes who sought perfection in unorthodox ways. Hughes's aircraft projects were given a Group V priority, the lowest rating.

According to the letter of intent between Kaiser and Hughes, Kaiser was left almost powerless to interfere: *'It is understood that the engineering and construction will be done under the direction of Howard Hughes.'* According to Kaiser, Jesse Jones had told him explicitly to keep hands off the project: *'I want it understood that Hughes has the responsibility and you do not interfere with him.'*

It was clear that Kaiser-Hughes was something less than a working entity. The flying boat was a Hughes Aircraft project from the beginning.

Hughes never attempted a design that was not unique compared to anything built by others. Hughes's ideas permeated every aircraft design, and the flying boat, which was his life for about two years, was no exception. He not only made all decisions relating to external appearance and size, but also was involved with powerplants, flight controls, and instrument panel design.

After the go-ahead to build the flying boat was signed in September

1942, Hughes got together with his engineers to decide its size, shape, and configuration. The first big question was whether it should be twin- or single-hulled.

James V. Martin, an eccentric aviation pioneer and creative genius, had earlier proposed a giant swooping gull-winged catamaran 'Twin Hull Martin Oceanplane' that had interested Winston Churchill and the British Admiralty. Churchill piqued the interest of President Roosevelt; and members of the Air Cargo Board (which Nelson had established under the auspices of the War Production Board that past May) strongly favoured such a design. Twin hulls would eliminate the extra drag and weight of wing tip floats and make it easier to beach the flying boat to discharge cargo in amphibious operations.

Loening expected that the big flying boat would be used for Pacific island amphibious operations to land in the sheltered water of lagoons and discharge troops and cargo from a bow ramp before backing off the beach with reversible-pitch propellers. Hughes himself had been impressed by the successful Savoia-Marchetti twin-hulled flying boats which had completed a mass flight from Italy to Brazil in 1931 under the command of Gen. Italo Balbo.

For weeks, every evening between 5:30 and 6 p.m., Hughes would arrive to argue the twin-hull-or-single-hull question. Kenneth Ridley would argue that twin hulls would weigh more for the same capacity but Hughes would disagree. Just as Ridley conceded that the twin hull design was right for the flying boat, Hughes opted for the single hull. This pleased the structural engineers, who had concluded that the twisting loads on the centre section between the two hulls would be excessive.

The single hull design would have its own problems, because the mammoth size Hughes chose for his design would require an overhang wingspan fifty percent greater than the Martin Mars, which would invite new torsional, wing flutter, vibration, deflection, and control problems never before tested.

Concept No.6 - Single hull and tail, side loading, but with eight engines.

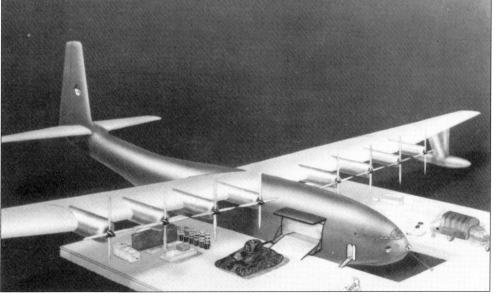

Getting Things Moving

Henry Kaiser finally had to call on President Roosevelt to provide the nudge to get the project going.

By now Hughes had completed all the necessary drawings and specifications for the flying cargo boat proposal. Contract negotiations were opened with the Defense Plant Corporation and the newly formed Kaiser-Hughes Corporation submitted their proposal with drawings and specifications on 17 and 18 October 1942. After the proposal was studied and the contract negotiated, it was signed on 16 November 1942 by Henry K. Kaiser, president, and Neil S. McCarthy, secretary for the Kaiser-Hughes Corporation.

It was a development contract for an entirely new type of aircraft. It was not for the purpose of mass-producing flying cargo ships. It called for three aircraft to be built for testing purposes. This fact was consistently omitted by most of the critics of the contract in later years. Neither the news media nor politicians ever seemed to give this item any credence.

The contract was very much like most developmental contracts issued by the government at that time. The document started off: *'This agreement made and entered into this 16th day of November, 1942, by and between Defense Plant Corporation (hereinafter referred to as 'Defense Corporation'), a corporation created by Reconstruction Finance Corporation pursuant to Section 5d of the Reconstruction Finance Corporation Act, as amended, to aid the Govemment of the United States (hereinafter sometimes called the 'Government'), in its National Defense Programme, and Kaiser-Hughes, a corporation*

Concept No.7 - the HK-1. single hull and tail, eight engines , nose loading and wing-floats on single struts. This is basically the final design, apart from refinements to the flight deck glazing and the row of 'portholes' aft of the flight deck that suggests that at the time a 'double-deck' aircraft may have been planned.

organised and existing under the laws of the State of California.

Whereas, Henry J. Kaiser and Howard Hughes have heretofore submitted to War Production Board, to Civil Aeronautics Administration, and to Defense Corporation designs (General Arrangement Drawing No. 55282, dated 10-l7-42, and weight and performance summary, dated 10-17-42 and 10-18-42) of cargo planes...'

There were pages of clauses regarding use of equipment, that the engines, propellers and instruments would be furnished by the government, who was to do what... One key phrases was that Kaiser-Hughes agreed that the engineering and construction work shall be done under direction of Howard Hughes.

'11 - Notwithstanding any other provision herein contained the maximum amount which Defense Corporation shall be required to expend hereunder shall not exceed Eighteen Million Dollars ($18,000,000), including the cost of installing, but not the cost of engines, propellers, or instruments furnished pursuant of Article 12 hereof. [this was the Government furnished equipment] *It is expressly understood that Kaiser-Hughes does not warrant that the project can be completed for Eighteen Million Dollars ($18,000,000) and in the event completion of the project requires further funds, nothing herein contained shall require Kaiser-Hughes to advance such funds.*

Clearly this shows that the cost of the project seems at first to be limited to $18 million, but further on, the Defense Plant Corporation acknowledges the costs could be much higher and that Kaiser-Hughes should not have to pay the difference. This was the provision in the contract for handling additional costs encountered in a project and it was the method used to provide for cost overruns. It is not difficult to understand that no company would accept a development contract at a fixed price. When you are dealing with experimental projects, it means that it has never been done and no one can predict how much it will cost

An artists impression of what the HK-1 would have looked like if it had entered service.

to do what's never been done.When the contract was let in 1942 the government envisioned an aircraft somewhat larger than the Martin Mars whose design gross weight was 145,000 pounds. Hughes finally settled on a design gross weight of 400,000 pounds. His choice resulted from a requirement to transport heavy artilliary weighing up to 80 tons. Ridley was asked to give a brief review of how they had arrived at an aircraft size of 400,000 pounds at a 4 May 1943 meeting in the NACA conference room in Washington. He stated that their objective had been to design the largest aircraft possible using eight of the largest engines currently in sight.

The interpretation of size arose again in a 21 October 1943 meeting attended by Hughes in Donald Nelson's office at the War Production Board. Grover Loening agreed that the objective was to build the largest possible load-carrying aircraft, but observed that the design was Hughes's own interpretation of 'the largest possible'. In Loening's opinion, this meant eight Pratt and Whitney 2800 engines in a machine of about 250,000 pounds gross weight, however '...*the engine Mr. Hughes has chosen, is not in production even yet and certainly was not at that time.*'

A Question of Size - and Material

Grover Loening was concerned about the planned machine's weight, because aircraft efficiency does not indefinitely increase with increasing size. Only the previous month he had written: *'It so happens that while larger airplanes get more efficient and faster as they increase in size due to the smaller relative drag of the body and relative weights of necessary items, like a pilot to fly them, there is also, due to the law of the cube, an increase in the relative weight of the wing structure with size. After the line of favorable decreases crosses the line of unfavorable increases, like wing weight, there is no further advantage to larger size aircraft.;* - a concept that Hughes was to later use in the Senate Hearings.

Hughes was concerned about another size-related problem. He felt that many new design features would need testing in flight, but that the team should not have to depend on flying the giant eight-engine boat every time some little gadget had to be checked. So in August 1943 Hughes's engineers proposed to Loening that they be allowed to build a near half-scale flying model of the HK-1 with eight air-cooled engines in a similar manner to what the British had done with some of their advanced flying boat designs.

Instead Loening recommended *'...that the Navy be required to loan to Kaiser-Hughes at once a PB2Y3 flying boat stripped of all military equipment and to be used by Mr. Hughes as his flying laboratory in connection with this contract.'*

Once the size, number of engines, and the single hull design had been decided upon, the Hughes team was ready to determine the specifics of shape.

Hughes was impressed with the work of the National Advisory Committee for Aeronautics' Research Center at Langley Field, Virginia. He had read an old NACA report by John B. Parkinson, then Assistant Chief of NACA's Hydrodynamics Division under Starr Truscott, that showed seaplane hulls drawn around streamlined bodies to minimise drag. After discussions with chief designer Rae E. Hopper it was decided to follow this approach.

Hopper and Kenneth Ridley, who ran the engineering division, called in preliminary design group engineers Roy Wendahl and Gene Blandford and

Buildings 15 North and 15 South are starting to be erected, while Building 16 is further advanced. There's still a long way to go.

told them to go to Langley to obtain NACA's latest design criteria for flying boats, as the company had had no previous experience with such aircraft. For some months the Hughesmen operated out of Parkinson's office and worked with the various NACA experts. Soon they were doing three-views of what was to become the biggest aircraft in the world.

Roy Wendahl drew up the hull three-views and sent them to Culver City as the basis for the preliminary design. There Hopper's design crew prepared sixteen profile views of the proposed aircraft which were then taken to Hughes at his house for approval.

After Hughes had made the final decisions on hull shape, his craftsmen at Culver City reproduced the beautiful and properly proportioned dynamic test model for tests at the NACA towing basin. Langley experts told Loening in February 1943 that this model was the finest job they had ever received from a contractor. It would be nearly a year before the hydrodynamic and aerodynamic tests were completed.

The hull design was only part of the work done at Langley. The wing section was recommended by NACA's Eastman N. Jacobs and use was made of Langley's low-turbulence wind tunnel. Carl Babberger believed that the result was one of the finest low-drag, high-lift sections that ever was put together.

Parkinson himself went on record saying that even unpublished ideas were applied to the nacelles so that they and their associated internal cooling systems represented the latest and the best in low-drag design. *'All in all, aerodynamically and hydrodynamically, the Hughes flying boat was very superior. We had high hopes of using it after the war as a research airplane simply because it did have such outstanding aerodynamic and hydrodynamic features. It had probably the largest Reynolds number I've ever heard of. We could have gotten a lot of aerodynamic data off of it. It had a finish just like a piano-probably the smoothest airplane that was ever built. That was one dividend from the Duramold plywood.'*

The model eventually used in the NACA towing tank and wind tunnel tests

The wind tunnel test model of the HK-1, complete with electrically-driven propellers.

was a powered dynamic test model with electric motors to drive the scale-size propellers. The wings and hull could be tested separately or together. Some wind tunnel tests were also conducted in Cal Tech's ten-foot tunnel by Dick Murrow using a 1/ 80th scale model with a four-foot wing span.

During this design period Hughes carried on an indirect dialogue with the NACA experts. He communicated in a kind of three-way conversation through Babberger, Roy Wendell, or other Hughes men at Langley. Often it was Rae Hopper who relayed Hughes's thoughts.

Occasionally Hughes disagreed with NACA. Like all pilots, he had strong ideas about everything concerning aircraft. A case in point was the design and location of the step in the bottom of the hull of the flying boat. Usually it was made too small, which gave rise to troubles during takeoffs and landings. So Hughes' team had the chance to say what they ought to be on the HK-1 and, according to all calculations, for the sixteen foot beam of the proposed boat came out as two feet, which was an enormous thing to have to drag through the air.

But even with a two-foot step the Hughes hull had the lowest induced drag over the equivalent streamlined body of any aircraft fuselage or body NACA had ever tested, and was probably the lowest drag seaplane hull that had ever been designed-simply because it had the area distribution of a streamlined body.

But it was the location of the step that Hughes was concerned about. He was certain that the step ought to be relatively far aft, so that when the aircraft touched on landing it would tend to put the nose down and kill the lift. To prove his point, he ran tests with his S-43, ballasting it to move the centre of gravity further and further ahead. Odekirk estimated that Hughes made six thousand takeoffs and landings on testing in a four-month period.

Buildings 15 North and 15 South starts to take shape as the cladding starts to go on.

Building 15 reaches completion and undergoes fitting out.
In this picture the interior is relatively dark, but it was later painted to reflect the light better as seen in the lower view, with some of the wing jigs in the centre of the picture, with the hull jig off to the left.
(both Simon Peters Collection)

Then a chain of events was set in motion that would lead to a tragic accident. Hughes had landed the Sikorsky on Lake Tahoe one day and then could not take off. No matter how he had rocked, pushed, and pulled, the flying boat would not break loose from the water. The engines simply had not had sufficient power to take off at Tahoe's elevation of 6,229 feet. Hughes had to wait for the wind to pick up and a chop to develop before the flying boat could take to the air. Hughes flew back to Culver City and ordered the original engines replaced with Wright G-205A engines. This increased the total power to 2,000 horsepower but also added weight and moved the centre of gravity forward. The change in the engines and the modification of the tail necessitated a flight test as a prerequisite to an updated airworthiness certificate before Hughes could complete a planned sale of the machine to the Army Corps of Engineers the following year.

C. W. 'Ted' Von Rosenberg, an engineering test pilot in the old Civil Aeronautics Authority Sixth Region headquarters at Santa Monica, was assigned to work with Hughes on the project. Flight tests of the Sikorsky with the new engines were completed by mid-May 1943. Only water tests remained, which were to be conducted on Lake Mead.

On the afternoon of 16 May 1943, the day before the water test, Von Rosenberg with copilot Ceco Cline, Chief of the Manufacturing Inspection Branch at the Santa Monica headquarters, flew the aircraft to Boulder City

The salvaged remains of Howard Hughes' Sikorsky S.43 N440 following its crash into Lake Mead. The aircraft was rebuilt and survives to this day.

on the south end of Lake Mead. Also in the aircraft was Hughes, flight test engineer Gene Blandford, flight engineer Dick Felt, and American Airlines pilot Howard West and his copilot. Tests began the morning of 17 May. The aircraft was configured in accordance with purchaser requirements. Radio equipment and a radio operator's position had been installed in the rear, so the rear ballast had been removed.

As the test began, Hughes took off from Boulder City Airport and flew to the north end of the lake for a landing on the water. Hughes was in the left seat, Von Rosenberg was in the right seat, and Felt, acting as flight engineer, was leaning through the cockpit door. Blandford sat in a jump seat on the pilot's side and Cline was in the upholstered chair at the navigator's table facing forward on the right side. Suddenly the nose pulled to the left and the plane started to skid sideways.

Hughes applied full right rudder. For a moment it looked as if he had regained control. Blandford, feeling the swerve, grabbed the nav table. Then all hell broke loose. By now, the aircraft was moving completely crossways. The right wing tip and float dug into the water, wracking the wing. Struts collapsed or were torn out. The whirling blades of the left propeller slashed through the hull right at Cline's jump seat. Finally the aircraft came to a stop. It was right side up but steeply nose down and listing about forty-five degrees to the right. Water gushed through the hole in the side of the hull slashed by the prop. The flying boat was sinking.

Hughes received a serious head injury in the crash, and it also dealt him a severe psychological blow. He had killed two men.

Work On The Big Boat Continues
As 1943 began, Hughes was still involved with preliminary design problems with the flying boat. WPB chairman Donald Nelson, who had taken the chance in approving the Kaiser-Hughes contract, grew worried by the lack of progress and instructed Grover Loening to make a surprise visit to the Hughes Aircraft Company on 10 January. It seems that Loening was not

The keel and vee-hull structure of the HK-1 takes shape 'in the stocks' inside building 15.

The wing went through many stages of construction and is seen here on its rear spar, leading edge pointing skywards. Of interest are the two track-mounted cranes used to move components around.

impressed. Among other things, he reported back that *'...it was difficult to get hold of Mr. Hughes, as he was busy on other matters.'*

Following Loening's suggestion that Hughes be forced to give full attention to the flying boat, Donald Nelson wrote to Hughes on 8 February: *'...The skill with which you complete this craft in a satisfactory enough state to be of distinct service is more important than any skill which might result in an ultimate perfection at the expense of time.'* The contents of this letter is contrary to a later assertion from Howard Hughes that appeared in *Newsweek* on 4 August 1947 in which he said that *'..I was specifically instructed that it was a research project and I was told to take my time'*

It is known that Hughes was rarely seen by most of the men who worked for him. His night-owl working habits, his almost obsessive desire for secrecy, and his involvement in a myriad of other simultaneous projects - plus a marked reluctance to delegate authority in matters of design - made working for Hughes a unique and often highly frustrating experience for his managers. Important decisions could be delayed for days if not weeks because Hughes's executives could not contact their elusive boss. In addition, he was always interfering - even remotely, he simply couldn't keep hands off and leave well alone.

Edward 'Eddie' Bern, the general manager of Hughes Aircraft Company for several months in the summer of 1943, was later to testify that he saw Hughes at the plant only once - but was also quick to add: *'it is my understanding that many times at two or three o'clock in the morning Howard would come in there with a ham sandwich and a glass of milk and work all night, but I did not see him.'*

During the first seven months of the project Hughes acted as general manager with the heads of the production and engineering departments

reporting directly to him. Henry Kaiser, after his unsuccessful first attempt at combining forces with Hughes, soon cut his losses and had nothing more to do with the project except to help out with a few personnel at the start and to drop in every few months for a hurried visit.

Such were the restrictions placed on using non-strategic materials, even the track-mounted cranes were made of wood!. Here one of the inner sections of the massive wing is in the process of being moved.

The problems with wood...

Major problems resulted from combining wood construction with giant size. The Hughes Aircraft Company had had previous successful experiences in wood construction; but this experience was not always directly applicable to building the flying boat.

'Aircraft Production' - the journal of the aviation industry in the United Kingdom - sent over an un-named writer at the invitation of Howard Hughes, and ran a series of articles on the HK-1, explaining in great detail the build process. It also revealed some figures as to the size of some of the components!

'Perhaps the most unusual feature of the aircraft, particularly in view of its great size, is that it is built almost entirely of wood. The reason for using wood was the critical shortage of light alloys at the time the preliminary negotiations on the contract for the aircraft were made. Regular aircraft birch, purchased in the northern part of the middle west-Wisconsin and Michigan, was selected because of past experiments with other aircraft and because of the strength to weight ratio. Also, spruce of the quality desired was difficult to obtain. Hughes Aircraft Company secured the licence to use the Duramold process of wood lamination.

Aircraft construction had never before required the handling of such large components, especially with wood design. Extreme precautions had to be taken and maintained to prevent the parts and assemblies from being affected by moisture and temperature changes. Existing methods and materials were found to be unsatisfactory, and many new wood-handling techniques were developed.

Engineers of the National Advisory Council of Aeronautics at Langley Field were consulted on the design of the hull. Their recommendations were used to give the best results in hydrodynamic characteristics. Several Hughes aircraft engineers made a trip to Langley Field for this purpose and the basic lines for the hull bottom were determined there. Preliminary designs of the hull were combined with the rest of the aircraft and a 1/16th scale dynamic powered model was built for tests in the NACA tank.

The results of these tests were found to be equal in good water characteristics to any model that had ever been tested and by comparison with other aircraft much better than the average.

Aerodynamic data to the Reynolds Number at which the Hughes Hercules would operate were not available, and NACA was again consulted in the determination of characteristics and the extrapolation of available data. A complete 1/20th scale model was' built and- tested at the tunnel of the Californian Institute of Technology.

Although birch is the principal constructional material, other woods, such as spruce, poplar, maple and some balsa, for fairings, is used in the aircraft.

All birch parts are laminated from veneer varying from 1/64th inch to 1/8th inch in thickness. Because of the great lengths of most parts, it was also necessary to scarf ends of veneers together before lamination. Birch billets for the wing spars were 6 inches by 8 inches by 90 feet long. Most spruce parts are made from solid stock; the longeron billets were 10 inchs by 10 inchs by 100 feet long and were built up from laminations 3/16th inch in thickness.

Because of the great size and quantities involved, it was necessary to design and build special scarfing cutters and presses. These scarf cutters were set to make scarf joints to tolerances never before attained in wood construction. In laying up billets, veneers were scarfed together selectively so that the variations in thicknesses of adjoining veneers would not

The huge two-piece wing during the early stages of construction. The wing is sitting vertically, on the rear spar. The size of it is noticable by the workers standing alongside it.

With the ribs and stringers in place, it was time to start covering the main surfaces of the aerofoils with a criss-cross pattern of veneers at 45 degrees.

culminate in glue lines of excessive thickness. As glue lines over a few thousandths of an inch in thickness are detrimental to long life, every glue line was carefully checked and parts showing excessive lines rejected.

It was found that the larger the part, the greater became the necessity of accurate moisture-control during manufacture. All wood was preconditioned before use, then gluing was done in buildings held at constant humidity and temperature between 72 degrees and 80 degrees. Different colours were added to the glue mix to allow identification at any stage.

Part of the skin on the tailplane and in surfaces is made of poplar plywood. Plywood skins with sharp or double curvatures were made in the company's

Lifting one of the wings from the assembly fixture for the installation of the trailing edge structure.

own press or by outside contractors to Hughes specifications which were more rigid than US Army-Navy specifications. Extensive use is made throughout the aircraft of plywood angle sections developed by the Hughes plant.

Hughes had purchased the partial rights - for aircraft above 20,000 pounds - to use the Duramold process for making molded plywood components from Sherman Fairchild in 1939. This process was ideally suited for making smoothly molded fighter-size aircraft but had not been devised for anything as big as Hughes had in mind.

He had previously bailed out Fairchild when money had been a problem. Hughes did so again and obtained the license to Duramold in the deal. Along with the Duramold process, Hughes acquired the services of Colonel Virginius E. 'Ginny' Clark and a young British wood expert, George A Allward.

The Hughes engineer assigned to work with Clark and Allward in the development of Duramold processes for Hughes Aircraft was mechanical engineer Louis Tribbett, who had started work for Hughes on 1 September 1939, in the old Western Air Express hangar at Burbank. Under Allward's supervision he tested all kinds of American timber and selected birch as the primary structural material.

During the war, Allward was hired away from Hughes by the Higgins Company while Tribbett was in the Pacific on military duty. Higgins was owned by Andrew Higgins based in New Orleans, Louisiana, United States. It is most famous for the design and production of the Higgins boat, an amphibious landing craft referred to as LCVP (landing craft, vehicles, personnel), which was used extensively in the Allied forces' D-Day Invasion of Normandy. Higgins also manufactured PT boats, and produced the first American airborne lifeboat, the model A-1.

Andrew Higgins also owned the New Orleans-based Higgins Lumber and Export Co., and Higgins Aircraft, which contracted to provide aircraft

One of the ailerons is moved by crane around the workshop. Of interest is the series of hull frames that made of the upper fuselage behind the flight deck of the HK-1 that are stacked against the rear wall of the building.

Slowly sub-components began to stack up awaiting final assembly. Here are the frames that created the flight-deck shape of the forward hull.

Although the later HK-1s would have an opening nose to allow cargo to be loaded, the prototype was fitted with a solid nose, as seen here awaiting fitment to the hull. *(both Aircraft Production - the journal of the aircraft manufacturing industry)*.

One of the forward hull frame assemblies. The bulkheads and frames were of all-birch construction.
(Aircraft Production - the journal of the aircraft manufacturing industry).

for the US military during World War Two. Hughes was furious and filed a lawsuit against Higgins.

Spruce, otherwise a good structural wood, could not take bolts well. But by laminating birch in both directions they could get the necessary grip of wood to bolt. They found, too, that for high stress applications birch in terms of weight was better. Additionally, birch made better molded plywood when glued and steam heated than spruce did. They also discovered that elaborate and costly jigs had to be devised and new glues and gluing procedures developed.

Intensive basic research in glues by two subcontractors resulted in some of the earliest practical solutions in the USA to the use of epoxy resins, which are chemical, thermosetting glues rather than organic glues. Further research and development resulted in special machinery and equipment for bonding, curing, and forming of high quality plywoods.

Three different types of epoxy resin glues were used: phenol formaldehyde resin cured by heating to 300°F, and became a commonly used glue for making plywood. It is cured at elevated temperature and pressure. Urea formaldehyde resin which cured at 70°F or above and had a low effective cost, low cure temperatures, resistance to microorganisms and abrasion, and light colour. It does not creep, and can be repaired with epoxy. However, it could rapidly deteriorate in hot, moist environments, releasing

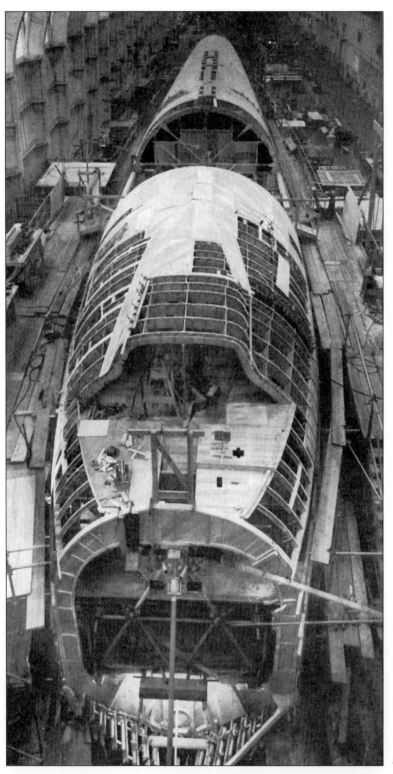

Looking down on the Hughes flying boat from a point forward and above the nose. Part of the flight deck has yet to be installed, as does the solid nose. *(Aircraft Production - the journal of the aircraft manufacturing industry)*.

Laying the woodwork
for the flight-deck area.
*(Aircraft Production -
the journal of the air-
craft manufacturing in-
dustry)*.

formaldehyde which was later discovered to be a carcinogen. Finally there was Resorcinol-formaldehyde resin type glue that became available after the project was well under way. This was a two-part, very strong and durable adhesive, resisting immersion in boiling water, mild acids, salt water, solvents, mold, fungus, ultraviolet light and the like. For many years, the Federal Aviation Administration (FAA) stated that *'Resorcinol is the only known adhesive recommended and approved for use in wooden aircraft structure and fully meets necessary strength and durability requirements'* for certificated aircraft.

Glue pressure for attaching the hull skin was provided by thousands of small nails driven by specially developed nailing guns. After the glue was cured the nails were removed by specially developed nail pullers. Approximately eight tons of nails were used.

The Duramold process is thought to have had its origins in 'Consuta', a revolutionary form of construction of watertight hulls for boats and marine aircraft, comprising four veneers of mahogany planking interleaved with waterproofed calico and stitched together with copper wire that dated back to the end of the 19th century. The technique was patented by Sam Saunders of Goring-on-Thames in the United Kingdom and was first used on the 1898 umpire's steam launch of the same name.

After opening the S. E. Saunders boatyard at East Cowes on the Isle of Wight, Sam Saunders further used the technique to build the crew and engine gondolas for HMA1 *Mayfly*, Britain's first airship. Later, the same technique was used to construct the hull of the Sopwith Bat Boat and a whole series of Saunders-Roe flying boats.

Just as Geoffrey de Havilland did when building the famous DH98 Mosquito, Howard Hughes sought out ways of solving the problems in constructing the large boat by looking in non-aviation directions. Assistance in solving tooling problems was provided by a number of consultants in the Los Angeles area woodworking firms who were doing subcontract work on subassemblies. The Weber Showcase & Fixture Company of Los Angeles, were contracted to manufacture bulkheads for the hull; the Edward I. Classen Manufacturing Company, Los Angeles, were contracted to manufacture stringers for the wing; the Modern Pattern & Foundry

Company, Los Angeles, were contracted to manufacture aluminum castings; Van Tuyle Engineering, Inc., Los Angeles, contracted to manufacture templates for parts manufacture; and Jasper Wood Products, of Jasper, Indiana, received the contract to manufacture plywood flat panels.

The flight deck area starts to take shape - the laminations of the hull frames being particularly noticeable, while inset; Howard Hughes and Odie Odekirk study the HK-1 blueprints in the loft of Building 15. *(Aircraft Production - the journal of the aircraft manufacturing industry).*

Aircraft Production further describes the aircraft: *'The hull is of all-wood construction with the exception of a few metal fittings. The skin bulkheads and frames are of all-birch construction while stringers are of composite birch and spruce design. The upper longeron is a 10 inch by 10 inch piece of spruce almost 100 feet in length. Parts such as bulkheads and frames are made by gluing the component parts together using clamps for pressure.*

The cargo floor was designed to carry a load of 125 lb./sq.ft. Supporting frames have been designed so that by laying the proper planks across the floor any heavy piece of equipment up to and including the Army's 60-ton tank, can be driven in under its own power and transported without dismantling. The bottom skin is ½ inch thick and is designed to take much higher bottom pressures than those used on smaller flying-boats.

Below the cargo deck the hull is divided into eighteen watertight compartments. If two-thirds of these were flooded, the ship would still remain afloat. The main fuel-tanks are housed in these compartments so that if a tank should leak the petrol will not spread to other parts.

The exterior of the boat is finished with a special process developed by the company. It consists of one coat of wood filler, one coat of sealer which acts as a cement for a coat of thin tissue paper placed over it, and two coats of spar varnish and one coat of aluminised spar varnish. Tissue paper was

The forward section undergoing precision sanding. The vee-section being worked on would, if this was a production aircraft, take the loading ramp for the nose camshell doors. *(Aircraft Production - the journal of the aircraft manufacturing industry).*

applied by regular paper hangers.

Over the passing years it has been claimed that some of this research conducted by Hughes Aircraft resulted in subsequent commercial applications. In the 1970s for example, an article appeared in the press quoting an un-named Hughes Tool Company executive as saying that '...*the big wooden bird became a hen that laid golden eggs. Everyone in the country, and quite a few firms outside, that make plywood have been paying licenses and royalties to Howard ever since.*'

This was disputed by a number of people, including Howard Hughes himself who said back in 1947 at the Senate Hearings that '*we have no patents which were intended to be used, we collected no royalty and I assure*

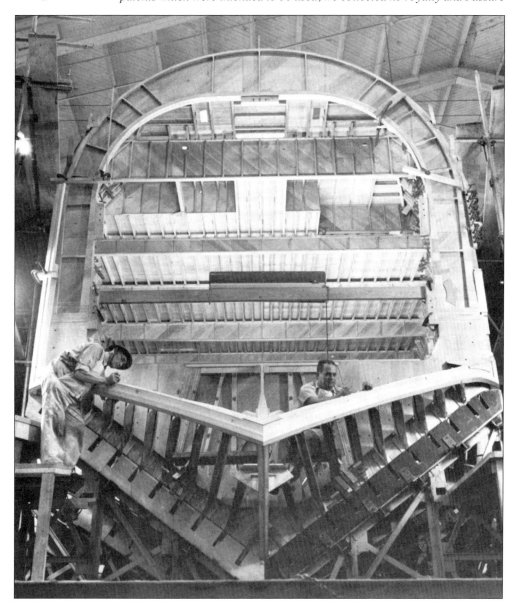

you we will not collect any and never had any intentions of collecting any.'

In 1972, John Ritchie, vice president and secretary of the American Plywood Association, said *'I'm sure that's not true* [the claim from the ToolCo Executive] *We would know it if it were.'*

Other researchers who made an extensive search of government patent files failed to unearth one single patent ever assigned to Hughes, Hughes Tool or Aircraft Companies, or to his key executives as a result of any HK-1 related work.

Hughes was lavish with equipment for Duramold, recalled Louis Tribbett. *'When we were building the D-2 we could have most anything we wanted. We went out and bought a steam tank, about ten feet in diameter, I think, and about forty feet long with a massive door on one end so we could move a whole section in. We'd put this glue and plywood in there in enormous molds and covered with rubber bags and then press it with steam pressure in order to 'cook' it.*

'We had to start from scratch with everything because the Fairchild data wasn't suitable for what we were doing, and when we tried to use some of the data from the government's Forest Products Laboratory they weren't worth a damn. It ended up that a lot of the research we did ended up in the new forest products book they copied from us.

'We didn't have any wood construction subcontractors until 1942. The Hughes company just went out and hired a bunch of Swedes and good woodworking men - some real fine people. The man we sent to Wisconsin to buy the selected woods was Gus Seidel, a personal friend of Hughes.'

After the contract for the flying boat was let, they had to expand the Duramold technology still further, according to Carl Babberger, Hughes's chief aerodynamicist: *'We built some fantastic types of Duramold plywood designed for the job it had to do - whether it was bearing or shear or tension - and developed special corner angles to replace glue blocks.*

The hull of the HK-1 takes shape 'in the stocks' at Culver City. When finished, the hull would be 220 feet long.

The interior of the flight deck, looking forward towards the pilots' positions. This picture was taken before the area was fitted out and all the control cables and services installed. *(Aircraft Production - the journal of the aircraft manufacturing industry).*

We demonstrated that glue blocks were a serious problem in wood construction because of differential expansion with and across the grain. I would say that the corner angle substitute for the glue block was the greatest development in plywood history. We went to real thin plies, maybe a hundredths of an inch thick, and then built up angles that had maybe ten or more plies with little fillets glued in at the angle.'

'The men wore white gloves in order not to put fingerprints on the laminations when they were building them up. A fingerprint was assumed to create a bum glue joint. We used the best woods and the best glues that were known.

'We also pioneered a special lofting technique. All templates were faired mathematically, machined out, and hand filed to a 1:16 scale. Then an enormous camera was used to project the templates to full scale. This produced continuous curves-beautiful things. After all, a spline is only as good as the eye of the man that uses it. But we could define everything, and did, to the thousandths of an inch. We were the first ones to try that'.

Birch was selected as the main structural material, not only because of the structural properties previously mentioned, and its good strength-weight ratio, but also because spruce of the quality desired was difficult to obtain. However, spruce, poplar, maple, and some balsa for fairings were also used.

The main structural material for the flying boat was built up by bonding several plies of birch veneer with a urea-formaldehyde glue. The bonds were formed under both heat and pressure, but some cold setting was used in certain cases.

Sharply curved surfaces, like the nose covering of the wings, were made using the usual Duramold process. But here the pieces were so large and the strategic material required for steel forms so hard and so expensive to obtain that Hughes experimented with 'Gunite'. This was a form of concrete that used dry cement and sand blown out of a hose with compressed air, that had water injected into the mix at the nozzle as it was released. The Gunite was blown against the negative surface of a form, thus creating a positive shape around which the Duramold could be produced. This is not as strange as it sounds, for De Havilland's used concrete formers for building the Mosquito that has been pioneered pre-war with their DH91 Albatross airliner.

This development of new tools, materials and methods was often by trial-and-error. Mistakes consumed time, material, and money. Progress was slow through spring 1943.

Engines, fuel and air

'Aircraft Production' went into great detail regarding powerplants and services. *'Eight air-cooled radial engines each developing 3,000hp will be installed on the leading edge of the wing.*

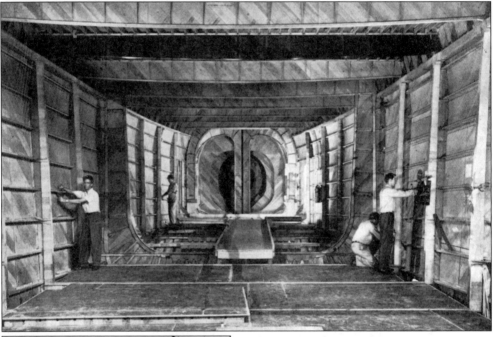

Above: the main cargo deck, looking forward to what in this aircraft is a solid nose, but if the HK-1 ever went into production would have been cam-shell doors. The floor of this area was designed so that with additional cross-planking, a sixty ton tank clould be driven in under its own power.

Below the floor of the main cargo deck, in the foreground in this picture and under the large outlined panels would have been fourteen, one thousand gallon fuel tanks like the one on the left. *(both Aircraft Production - the journal of the aircraft manufacturing industry).*

Through a passage in front of the front spar, and then through the nacelles and fire-wall doors, the flight mechanic may inspect and make minor repairs and adjustments on each engine. All fuel, oil, hydraulic and pneumatic plumbing is taken along the spar providing for quick inspection. Nacelle structures are all-wood and are mounted on the front spar by welded steel-tube adapters.

All cowling and other structure forward of the firewalls are of metal. Air for the oil-coolers and carburettors is admitted through underslung intakes on the nacelles. The intake fairings also enclose the oil-coolers.

A small oil-tank sufficient to supply the engine for a reasonable time is located in each nacelle and may be refilled at any time from a central oil reservoir by a semi-automatic control system. With the object of converting as much of the exhaust-gas energy into thrust as possible the engines are fitted with 'jet stacks.'

Hydromatic, full-feathering, four-blade airscrews, 17 feet 2 inches in diameter adjusted to produce reverse thrust. This feature will give greater manoeuvrability in water and assist in reducing speed after landing if necessary. A blower fan is fitted to the airscrew spinners to provide engine cooling under reverse-thrust and other critical cooling conditions, such as taxying.

Fourteen fuel-tanks, each of over 1,000 gallon capacity, store fuel in the hull. Two transfer pumps supply fuel from these tanks to a service tank in each wing. Each tank supplies fuel to four engines selected to give a lateral balance of power in the event of one tank, its booster pump or system failing. In addition to this regular fuel-supply system, a complete and separate system providing enough fuel at all engines for maximum power at the

The view looking through one of the eight engine nacelles undergoing finishing. *(Aircraft Production - the journal of the aircraft manufacturing industry).*

required pressure at the carburettor from the hull storage-tanks is immediately available for emergency use. An interesting item in connection with this system is that the fuel lines range in size from one inch to three inches in diameter. Another item is the fact that wing deflection relative to the plumbing made necessary slip joints in all fuel lines and boating fairleads. Throughout the aircraft, fitting sizes and equipment of adequate capacity were generally not available and required special designs.

The control of eight engines, all of them located at very great distances from the flight engineer and pilots was a serious problem considering the delicacy of adjustment and accuracy of motion required by engine controls.

The engineers responsible for this task went to railway practice and adapted the Pneudyne compressed-air control system, which had a long record of reliability in brake control.

Pneudynes, as the name suggests, are pneumatic control devices for the precise transmission of very small motions at long distances. Any position

One of the eight engine nacelles is installed onto the wing. *(Aircraft Production - the journal of the aircraft manufacturing industry).*

One of the air-intaker scoops inverted on the assembly jig.
(Aircraft Production - the journal of the aircraft manufacturing industry).

of the lever on the transmitting Pneudyne is maintained accurately on the receiving Pneudyne, for indefinite periods of time, even if the atmospheric temperature and pressure change during flight, provided of course that compressed air is available.

To secure the supply of compressed air, the aircraft is equipped with electrically-driven compressors but reliance is not placed upon them during flight.

These compressors provide the air supply and are operated automatically to maintain desired pressure in the air bottles. In event of compressor failure the air-bottles have sufficient capacity to outlast a normal flight.

The pneumatic system is completely enclosed, requiring only a small amount of make-up air and is equipped with moisture-absorbing devices to safeguard against freezing. All engine controls are actuated by the Pneudynes from an impressive flight engineer's station, the pilots controlling only the engine throttles. The pneumatic engine-control system solved the difficult task of transmitting delicate motions and adjustments to long distances against structural deflections, temperature changes and humidity variations of a large aircraft.

Owing to their immense size, the control surfaces cannot be operated by manual effort, but are actuated by a hydraulic telecontrol system which transmits the pilot's controlling motions, and amplifies his efforts many times.

Operating power is supplied by electrically- driven, high- pressure, hydraulic pumps which supply oil to sensitive relay valves actuated by the pilot. The valves are operated by conventional controls. High-pressure oil is metered by the transmitting relay-valves and directed by conduits to the control-surface stations. Here, the impulses are intercepted by receiving relay-valves, which in turn admit oil into power cylinders containing pistons connected to the control surfaces.

This hydraulic telecontrol system is not only capable of transmitting the

The hull slowly came together with the diagonal planking in place...

pilot's motions to the control surfaces, but being reversible, it also signals gust forces back to the pilot, thereby anticipating the impending change in the altitude of the aircraft while it is in flight.

To afford a greater measure of safety during flight, each control surface is operated by two completely independent self-contained and separately sufficient telecontrol systems supplied with electric power from separate generators, In case of the failure of one of the hydraulic systems, the remaining one is capable of taking control automatically and operating the control surface under all conditions of flight.

As has been mentioned, the pilot's and co-pilots flight controls in the cockpit are of conventional type with a column and wheel for the elevator and ailerons. Rudder pedals have a motion parallel to the floor, similar to the old-type rudder-bar. This type proved to be lighter and simpler than any other.

The trimming tabs are operated electrically from switches in the cockpit. The tabs are actually operated by electric linear actuators attached to them.

...that was then sanded, covered and then sealed before being finished. *(Aircraft Production - the journal of the aircraft manufacturing industry).*

Timber or Metal?

In later years Hughes partially blamed delays in building the giant flying boat on the requirement that he build it of wood - classed as a non-strategic material.

However, government records show that very early in the life of the contract, Hughes was given the option of switching to metal and made no effort to do so. By all accounts, it seems that the effort he had expended, the progress the Hughes team had made in pioneering methods of wood construction, and Hughes's attraction to the smooth external finish of the Duramold plywood made him reluctant to change materials after the start of the project.

The opening shot in the wood-versus-metal controversy was fired by Grover Loening on 3 April 1943, in commenting on the preliminary general specifications for the airplane. *'...The use of wood, because of strategic shortages, is not now needed...'* and he recommended that the wing be designed for metal!

A few weeks later, on 4 May, William Burden, who as Special Aviation Assistant to the Secretary of Commerce represented the Defense Plant Corporation, chaired a meeting in the NACA conference room to review

The pressure tanks used for the Duramold process. The largest was twelve feet in diameter and thirty feet long. *(Aircraft Production - the journal of the aircraft manufacturing industry).*

Wood sections were treated with a special resin and then subjected to steam heat and pressure inside the tanks.

the progress on the HK-1. Henry Kaiser was there; Hughes was not. Representing Hughes were Kenneth Ridley, his chief engineer, and Edward Bern, then general manager of the Hughes Aircraft Cargo Division. During the meeting Burden turned to Ridley and asked, '...are you satisfied with the wooden construction that has been specified for this work?'

Ridley, on behalf of Hughes, replied that he would not ask for any change in material: '...I do not feel that there would be any advantage whatever in going to metal construction. On the contrary, I think it would take longer to build, cost more, and might even be heavier.'

Captain L S Stevens, representing the Navy went on record stating that the Bureau of Aeronautics was already on record as being against wood construction for the project. The Navy felt the design was very good, but was skeptical of wood and the time of construction.

Rumours circulated that even if the engineering department may have preferred metal, Hughes may well have thought that wood was better. Many thought that Hughes could well have feared that if he switched to metal at this stage then the government may have replied 'Well, you're going to be delayed some more, the war will be over, we'll just cancel the project.'

From 2 August 1943 Loening, Burden, and Dr. George W. Lewis of NACA, under whose direction NACA had helped with the design of hull and wings and conducted aerodynamic and hydrodynamic testing, visited the Culver City plant.

Loening reported: 'The general progress of the Kaiser-Hughes project can be reported on as satisfactory. It is, of course, difficult to appraise work of this character, but the progress on the engineering seems particularly noteworthy. The design of the KH-1 aircraft, aerodynamically and hydrodynamically, appears excellent. It is in no sense a freak, but is, on the other hand, a very fine and graceful layout in a fairly conservative but thoroughly modern type of aircraft in which giant size alone is the principal new contribution.'

Then Loening again raised his concerns about using wood for the construction. *'Several times in the past, your consultant has called attention to the serious question involved in building the huge 320-foot span wing of this design out of wood. Originally this was mandatory upon Hughes, because it was and still is distinctly understood that this project must not 'raid' strategic materials from military aircraft production. However, since that time the situation has reversed itself in that wood, particularly birch veneer, is becoming very scarce, and metal is becoming a great deal easier. For this reason, the question of developing a metal wing for this design was again taken up in detail with Mr. Hughes. In the discussions on this subject, Mr. Hughes finally expressed agreement and said that he would be entirely willing to proceed at once with the metal wing.'*

Loening concluded his report: *'Since there is growing very quickly now an appreciation of what is coming in Air Power development in the way of invasion task forces that move entirely by air, the Kaiser-Hughes development is daily assuming more sense and importance, despite the skepticism in its early days .... In general, the Kaiser-Hughes organization is showing imagination, initiative and lots of push in connection with this work. They are meeting with all kinds of troubles, but they are facing them properly, and it is again urged that this development be pushed to the utmost.'*

Workers toil on the trailing edge of the port wing of the HK-1. The cameraman is standing in the cut-out where the wing eventually would fit into and across the fuselage. *(Aircraft Production - the journal of the aircraft manufacturing industry).*

Three weeks later Loening admitted that he had been conned. LeDuc, Bern, and half a dozen others resigned after Hughes had represented to Loening *'...that his management and organization were in running order.'* Loening admitted to Nelson in a memo of his great concern that the finishing of the stabiliser indicated an overweight problem *'...far beyond what would be permissible and of a nature requiring immediate review of the project'.*

Nelson sent Loening on a intensive investigation of the situation at Culver City. In his report of this 20-22 September visit, Loening stated that although he saw difficulties in the manufacturing organisation, he thought these an

The wing, having been turned horizontal ready to have the engine nacelles installed. *(Aircraft Production - the journal of the aircraft manufacturing industry).*

indictment of Hughes's management. *'As a matter of fact, conditions in the fabricating end of the plant were generally reported as greatly improved since the recent resignations of Messrs. LeDuc, Bern and others. There is not, therefore, any particular necessity for appraising the progress of this contract and the future plans for it in the light of management difficulties, but only in the light of technical considerations of the design itself. The most serious thing that is taking extra time… is that the method of construction has not yet been wholly devised and actually, the government is financing an experimental development of a new wooden construction method-at a time when it thought it was financing the development of a giant aircraft built on known structural fabricating methods'.*

Due to the outsized members in the huge wing that were to take the stresses, a great deal of work went into making up the billets from which these members were shaped. These billets might be composed of as many as fifty laminations all pressed together with a very fine glue-line tolerance. If the glue was too heavy or too thick, the strength was impaired and the weight increased. If the glue was too thin, a good bond was not obtained. There were frequent rejections of components due to glue lines indicating an insufficient bond, and Loening considered that there was no way to inspect the actual internal quality of the bond. This proved to be a major criticism regarding wooden construction of this type and size, and Loening jumped on it.

'Unlike the ability we have in metal with magnaflux and x-ray devices to determine the character of metal joints, these wooden joints must be taken on faith and on their external inspection only. Yet structural loads including vibrations, flutters, etc., and finally, weathering effects, will definitely strain these joints with forces other than shears - in which they are manifestly strong - and thus give to the entire structure an element of unreliability that is definitely 'scary' for aircraft'.

Loening concluded the report of his late September visit by stating '...*It is the details of construction material used and likelihood of serious overweight, the miscalculating of time needed and cost, the choice of engines, and consequently the practical availability of the aircraft for this war that warrant a very serious review of the desirability of this project from here on*'.

Similar doubts were expressed by the Chiefs of Aircraft Engineering, Flight Engineering and Factory Inspections Divisions who sent a letter to the Director of Safety Regulations, Civil Aviation Authority on 7 September. They stated that in view of the size of the aircraft, the materials used in its construction, and the difficulties encountered, '...*it is our considered opinion that further work on this project would have very doubtful value if the construction of the aircraft is to be considered a war project*'.

Clarence Selberg, one of Hughes's aircraft inspectors, disagreed: '*In quality standards and workmanship, the flying boat is perhaps the best wood structured aircraft ever built. Hughes was very meticulous regarding stress of materials, quality of workmanship, and appearance. Wood parts to be assembled were mated to each other at the glue bond area to a maximum of .003 inches gap.*

Blue chalking on one of the glue bond area faces was checked for transfer to the other. Where possible, feeler gauges were also used to insure precision fit. And a method of testing soundness of glue joints was developed using sonic as well as x-ray methods.'

The review of the flying boat project recommended by Loening began on 4 October 1943 with a meeting of the Aircraft Production Board, a subordinate activity of the WPB. This meeting was chaired by Charles E. Wilson, the WPB's Executive Vice Chairman. Minutes were taken by Loening.

One of the HK-1s flaps comes to completion before covering. *(Aircraft Production - the journal of the aircraft manufacturing industry).*

After detailed review and discussion of Loening's recent reports, '...*It was the decision of the Board that the contract should be cancelled, inasmuch as the project offers no useful contribution to the War effort, and the facility is required for production of important military types.*' Almost as an attempt to soften the blow of the cancellation, Brig. Gen. B. E. 'Benny' Meyers, who represented the Army Air Force on the Board, advised that they were '*..ready to place a contract for the Hughes photographic type airplane immediately upon availability of the facility for the purpose*'

Referred to as 'engine nacelles', the terminology is not strictly correct - they are more 'engine support structures'. The covers over each are to protect from worker-damage, and was typical of Hughes' attention to detail. *(Aircraft Production - the journal of the aircraft manufacturing industry).*

Following this Board meeting Wilson chaired a second meeting - this one attended by William Burden, Assistant Secretary of Commerce for Air; Grover Loening, Aircraft Consultant, WPB; C. L. Stanton, Civil Aeronautics Authority; Dr. George W. Lewis, Director of Aeronautical Research, NACA; and A. A. Vollmecke, Civil Aeronautics Authority. After the situation and the relevant factors were reviewed, it was understood that unless the Services objected the project would be cancelled.

To give Kaiser and Hughes ample opportunity to answer all the points that had been raised at these meetings, they were invited to Washington for three days of conferences. The first meeting, chaired by Grover Loening was held in Nelson's office at the War Production Board on 21 October. Loening opened with a brief review of the history of the project and the altered circumstances that led to the decision on 4 October to abandon the project. Then Burden and Vollmecke expressed their opposition to wood construction and Loening quoted Dr. Lewis's opposition.

According to the minutes, Hughes listened and then said, '*...Only one wood plane, that in which Knute Rockne was killed, has cracked up as a result of structural defects. My own planes have all met the most strenuous tests. It was in a plane with wood wings, for example, that I made the record* *flight from Los Angeles to New York eight years ago at an average speed of 325 miles per hour.'*

'The 'Knute Rockne' aircraft Hughes was referring to was a wooden-winged Fokker F.10 tri-motor airliner of TWA which crashed into the Kansas prairie, killing popular sports hero Knute Rockne and seven others on 31 March 1931. Rockne was an American football player and coach, both at the University of Notre Dame and regarded as one of the greatest coaches in college football history.

The accident was arguably caused by the composition of the aircraft. At

The aileron - closest to the camera - and flaps are fitted to the trailing edge of the wing. The cut-out in the aileron is for the trim-tab. *(Aircraft Production - the journal of the aircraft manufacturing industry).*

that time, the wings of Fokker Trimotors were manufactured out of wood laminate; in this instance, moisture had leaked into the interior of one wing over a period and had weakened the glue bonding the structural members that prevented the wing from fluttering in flight. One spar finally failed; the wing developed uncontrolled flutter and separated from the aircraft. Driven by the public feeling for Rockne, the crash story played out at length in nearly all of the nation's newspapers, and gradually evolved into a demanding public inquiry into the causes and circumstances of the crash.

Building 15 - both North and South - close to completion at Culver City towards the end of 1943. Its dimensions were enormous; 750 feet long, 250 feet wide and over 100 feet high. Off to the left is the grass runway.

William Burden made it clear that in contrast to British practice, the trend in the USA had been away from wood.

Hughes retorted that he did not believe that wood had been dropped for metal entirely on the grounds that wood is an inferior material. *'Other factors played a part in the switch from wood to metal. One, the public was easily led to believe that metal is safer than wood. Two, the suppliers of light metal are more aggressive, and perhaps more willing to cooperate with the plane manufacturers, than are the suppliers of wood. And three, it is likely that financial tie-ups between the producers of planes and light metals have had something to do with the switch. If the Army and Navy have given up wood, as Mr. Burden says, I think the decision might have been influenced by the fact that the builders for the Army and Navy have not always been skilled in wood construction. And the recent development of phenolic glues places wood construction in an entirely different light.'*

Loening asked Hughes, *'Do you consider wood satisfactory in the construction of seaplanes as well as land planes?'*

Hughes: *'With phenolic glues wood might even be considered superior to metal for seaplanes.'*

Loening raised the question of water soakage, and in the discussion he and Hughes seemed to be far apart in their estimates of the amount of water the HK-1 would absorb.

Burden, speaking for the Defense Plant Corporation, said, *'I very much doubt that this plane can be completed before the end of 1944.'*

Again Hughes disagreed. *'I'm certain that we can deliver the first plane within one year, and as a guarantee, my company will forfeit a half million dollars if that schedule cannot be met.'*

When Loening commented again on the excellence of the design and stressed that he strongly favoured redoing it in metal, Hughes retorted, *'I see no reason why wood is not as good today as it was a year ago. But if the War Production Board now takes the view that wood is unsatisfactory, it follows that the War Production Board was ill-advised to authorize the project in the first instance.'*

By now tempers were getting somewhat frayed. Loening: *'The circumstances under which the original decision was made have been radically altered, and in any case the question under consideration is not whether the original decision was a wise one but whether the project should now be continued.'*

Kaiser suggested that one machine be constructed of wood and one of metal so that they could be tested and compared, but Loening objected. *'I favor proceeding in metal only, but it's not certain that Hughes's design staff is qualified to work in metal.'*

Now it was Hughes turn to become annoyed: *'I can assure you that my staff is perfectly capable of switching to metal'.*

The meeting adjourned with no formal action taken. But the group appeared generally agreed on the desirability of salvaging the design.

After the meeting, Kaiser and Hughes visited Wilson. Both men believed in going to the top. They found Wilson cordial in attitude and soothing in manner. Wilson strove to understand their side of the story, for he knew full

The wing was initally built with cut-outs left for the later installation of the engine nacelles. *(Aircraft Production - the journal of the aircraft manufacturing industry).*

well the power of the Kaiser-Hughes financial, political, and public relations clout, and he knew that he could not afford to be wrong.

An excellent impression of what is meant by a 320 foot wingspan! The port and starboard wings are in dihedral position, ready for test mating. *(Aircraft Production - the journal of the aircraft manufacturing industry).*

Both Kaiser and Hughes were impassioned, persuasive and persistent. Hughes was experienced in fashioning arguments of great plausibility. He had answers for everything. In turn Kaiser spoke of how Douglas, Martin, and the other established aircraft manufacturers had been against him from the start. With just a hint of paranoia in his voice he suggested that a monopolistic conspiracy had been against their project from the very start and at every step of the way. As a result of their pleas, they managed to extract a very significant concession from Wilson. According to a memo by a Miss McCray of the WPB executive office, *'Mr. Wilson made a verbal agreement with Mr. Hughes when he was here and is not going to put it in writing. He has a modified 30-day extension and says he will have some part of the plane completed at the end of that time . He plans to talk to Mr. Loening about the letter and agreement. The Army and Navy are in agreement on the procedure and have promised to loan experts to go out and examine the operation there.'* This typed memo, dated 3 November 1943, was stapled to the minutes of the 21 October meeting.

Wilson advised the Air Production Board at their 25 October meeting that *'...the cancellation of the Defense Plant Corporation contract for the Kaiser-Hughes cargo airplanes has been deferred for thirty days, after which a reappraisal of the project will be made from an engineering point of view.'*

The situation was clarified at the Aircraft Production Board meeting on 1 November. *'The final decision is that the Kaiser-Hughes Company will be allowed approximately 30 days to static test a number of subassemblies, at which time an examining board of engineering experts will be sent to the Kaiser-Hughes factory to determine the engineering feasibility of the project.'*

The Examining Board will make its report to the Aircraft Production Board, where the final decision will be made relative to the disposition of the project'

Grover Loening was very unhappy with all the avoidance of making a firm decision. Before the Engineering Examining Board had even begun its work, CAA inspectors and engineers reported that static tests of the flying boat's stabiliser conducted prior to the 1 December deadline had been unsuccessful: *'The test stabilizer was completed on 23 October. The first test was started and discontinued on 13 November due to failure at 55 percent of the required design loading. The principal failures involved were glue joints (skin-stringers). Several of these glue joints were reinforced prior to the next test'*.

'On 19/20 November the stabilizer was successfully tested to limit loading (67% design loading) for the maneuvering up load. Prior to the next test, however, additional reinforcements to the structure were added.
'On 26 November tests for the second condition (balancing-down load) were discontinued because of glue failure at 50% of the required ultimate loading. Following this test, all stringer to skin attachments that had not already been reinforced were reinforced'.

Three days later, on 29 November Loening sent a memo to Wilson saying that with the expiration of the 30-day period agreed upon, the unsuccessful tests should be brought to the attention of the Aircraft Production Board at once. In his last paragraph he said: *'The recommendations made to you by Dr. Lewis of the National Advisory Committee for Aeronautics, Mr. Burden of the Department of Commerce, Mr. Vollmecke of the Civil Aeronautics Administration, and myself, for the immediate cancellation of this wooden construction, appear to be further reinforced by the lack of successful results in the last few weeks. In view of the expert knowledge of the National Advisory Committee representatives and the inspectors and engineers of the Civil Aeronautics Administration, additional engineering investigations would appear to involve an*

Inside the fuselage, looking aft. *(Aircraft Production - the journal of the aircraft manufacturing industry).*

Termed by the Hughes organisation as 'pontoons', the wing floats are seen here under construction. *(Aircraft Production - the journal of the aircraft manufacturing industry).*

undesirable delay in stopping the expenditures of public monies'.

Loening thought that the static testing of the tail was not nearly as important as testing for flutter, because the possibility of flutter is very serious in wooden aircraft. *'Due to the peculiar action of wood in various humidity conditions, one day it will be perfectly okay, and the next day its rigidity will be quite different, and that has always been one of the reasons why aircraft engineers would like to leave wood and go to metal, because they can count on it more surely. The stress can be figured more accurately'.*

The 29 November meeting of the Aircraft Production Board considered Loening's recommendation that the wooden construction be immediately discontinued and that the Engineering Examining Board investigation be cancelled. However, the Board concluded that '..*in view of the War Production Board's agreement with the Kaiser-Hughes Company to appoint an*

The completed wing floats before attachment to the struts. *(Aircraft Production - the journal of the aircraft manufacturing industry).*

'*Engineering Examining Board*' to determine the engineering feasibility of the particular type airplane, the investigation shall proceed as originally planned'

Wilson explained the rationale behind his thinking to Loening in a memo on 8 December: ...'*It would appear to me that if we follow the procedure on which we obtained a substantial agreement, our case will be stronger in combatting the almost certain pressure from various outside groups if and when cancellation is finally effected .... Failure to do so will give Messrs. Hughes and Kaiser material on which to pin a story of violation of agreement, persecution, etc. Furthermore, as nearly 20 million dollars of the taxpayers' money has been spent on a single model, it seems to me that, even if the final decision is to cancel, the development experts should make such a decision on the ground, in view of the magnitude of the proposition.*'

Due to the gluing difficulties revealed by the structural tests, the CAA arranged for a glue expert from the Forest Products Laboratory to consult with the Hughes Company on these problems. Since October the Forest Products Laboratory had been conducting their own intensive tests of a new and very promising resorcinal glue that could be applied cold.

By this time Loening was certain that the project had little chance of making a wartime contribution. On 27 January 1944 he asked Dr. George Lewis of NACA whether or not the general progress and prestige of American aviation from a postwar technical standpoint would be adversely affected by cancellation of the construction in wood of the HK-1.

Dr. Lewis replied that '*This project of constructing the HK-1 flying boat of wood serves no useful purpose in the interest of advancing the American aviation art*'.

The rear section being assembled. *(Aircraft Production - the journal of the aircraft manufacturing industry).*

By now the Engineering Examining Board under Dr. E. P. Warner of the CAA had completed its work. Someone extracted some lessons from the memo report on February 3, 1944:

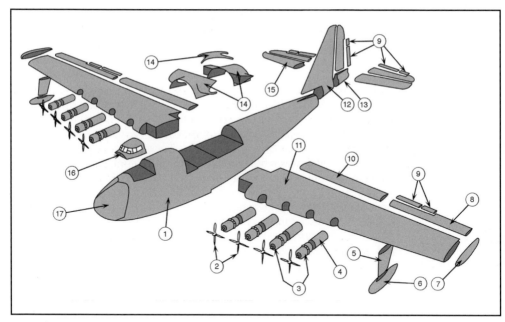

1 - Hull Structure
2 - Propellers
3 - Powerplant
4 - Nacelle Structure
5 - Float Support
6 - Float
7 - Wing Tip
8 - Aileron
9 - Trim Tabs
10 - Flap Assembly
11 - Wing Assembly
12 - Vertical Fin
13 - Tail Cone
14 - Wing/Hull Fairing
15 - Horizontal Stabiliser
16 - Flight Deck Enclosure.
17 - Nose Doors (on production examples)

' *The HK-1 design is of such a conservative and normal pattern that its performance can be very accurately predicted. This has been done by Dr. Warner's committee and the indicated performances are definitely of an inferior nature due to this excessive over-weight. It is unnecessary to complete the plane in order to verify these performances .... A most valuable lesson learned from the HK-1 development is a clear indication that efficiency of cargo carrying aircraft does not increase indefinitely with size in any such degree as had been generally considered likely before this work was undertaken. The over-weight difficulties on the Hughes design emphasize that the size has gotten too big and into a white elephant class where efficiency in load carrying becomes sacrificed to an excessive structural weight empty of the aircraft itself. The indications are that the most efficient size would be about half way between the Martin Mars and the Hughes design, even if the latter were to be built in metal'*.

Loening and Nelson visited the Martin plant and inspected the Martin Mars programme. It seems that they liked what they saw for Loening added an additional new argument to his recommendation to cancel the Kaiser-Hughes contract: *'The successful development and demonstration of the Mars flying boat makes further prosecution of the Kaiser-Hughes boat of no vital interest to WPB'.*

Enter The Truman Committee

It looked as though contract cancellation was inevitable, but Hughes was not ready to give in yet. Not long after the Warner report was issued, Kaiser and Hughes began telephoning Donald Nelson. Nelson had not even had time to read the 65-page report, which he had planned to read that weekend, when he received the first phone call from Kaiser on Saturday morning of 29 January.

Kaiser's pitch was emotional for a supposedly pragmatic businessman whom one would expect to be more factual and rational. He alleged to Nelson that Grover Loening had double-crossed him in connection with flying boat. Kaiser threatened to go public and say what he thought about him and what he (Loening) had done to him. Kaiser is supposed to have said *'I will get the whole goddam thing out in the open, including Donald Douglas and Martin and how they treated me and how they took evidence and then destroyed the evidence!'*.

Nelson cautioned Kaiser, advising him not to go off half-cocked until he got the facts. *'The Truman Committee has got hold of this and they are at work on it. And we have got to do the right thing regardless of passion or prejudice or anything else. I am not going to let any of that influence me, and it must not influence you'*.

This the first time that 'The Truman Committee', formally known as the Senate Special Committee to Investigate the National Defense Program comes into the story of the Hercules, but it would play a very great part in it later on. It was a United States Congressional investigative body headed by Senator Harry S. Truman. The bipartisan special committee was formed in March 1941 to find and correct problems in US war production — problems with waste, inefficiency and war profiteering. The Truman

Senator Harry S. Truman and others at a meeting of the Senate Special Committee to Investigate the National Defense Program (the so-called Truman Committee), held in the Senate Caucus room in 1943. In the foreground from left to right: Senator James M. Mead, associate counsel Charles P. Clark, Senator Harry S. Truman, Senator Ralph Owen Brewster, Senator Joseph H. Ball, and Senator Gerald P. Nye.

Committee proved to be one of the most successful investigative efforts ever mounted by the US government: an initial budget of $15,000 was expanded over three years to $360,000 to save an estimated $10–15 billion in military spending, and thousands of lives of US servicemen. Chairing the committee helped Truman make a name for himself beyond his political machine origins, and was a major factor in the decision to nominate him as vice president, which would propel him to the presidency following the death of Franklin D. Roosevelt.

Truman stepped down from leadership of the committee in August 1944, to concentrate on running for vice president in that year's presidential election. From 1941 until its official end in 1948, the Truman Committee held 432 public hearings, listened to 1,798 witnesses and published almost 2,000 pages of reports.

A few hours after Kaiser had spoken with Nelson, Hughes telephoned and asked if he could call Nelson at home during the weekend. Nelson gave him a suggested time.

On Monday 31 January Nelson received a five-page telegram from Kaiser. Nelson had said he would read the engineering report of 15 January over the weekend and Kaiser already wanted to know what he thought. *'You probably have now reached the same conclusion as our engineers have that there is nothing in the report on which to base the stoppage of this important development program. You will note the committee made no recommendations as to whether the project should be cancelled and that all their recommendations are based upon continuance of the project.'*

Despite the fact that the government had always envisioned the fastest possible development of aircraft to carry wartime cargoes, Clearly from Kaiser's telegram he had assumed that the flying boat project had been experimental from the beginning. *'The Kaiser-Hughes flying boat is purely a development project. It was conceived as such and recognized as purely development prototype project and is now in an advanced state of completion. In furtherance of air cargo with courage and vision in 1942 you courageously authorized the contract with Kaiser-Hughes this prototype for a large flying boat as an experimental project. We accepted it as a no-fee contract on that basis and I was forced to forego my plan of producing in quantity the Mars type plane on the insistent promise that development was an important essential. I can hardly conceive now when the plane is near completion and over ten or twelve millions have been spent that we must hesitate and fall.'*

Kaiser ended his telegram by saying that Howard Hughes wanted the Aircraft Production Board to hear him analyze and present his answers to the 15 January report. *'Certainly he should be accorded this hearing'* concluded Kaiser.

On 16 February Jesse Jones, the Secretary of Commerce acknowledged Nelson's letter of the 11th requesting immediate cancellation of the existing contract between Defense Plant Corporation and Kaiser-Hughes. Further, as Nelson had requested, DPC had invited a prompt and reasonably detailed proposal from Kaiser-Hughes for the development of the HK-1 design in metal.

The next morning Nelson appears to have had second thoughts and telephoned Jones or his Assistant Secretary, W. L. Clayton, to request that Hughes be allowed to continue work for a few days on the wooden flying boat until receipt of his new proposal.

It did not take Hughes long to hear of developments. That same day he called Nelson: *'I talked to Will Clayton. He said that they would continue the project for a few days pending your consideration of our proposal.'*

Hughes explained that in his bid he had inserted a paragraph stating that the job will be done at cost and that if the cost exceeded the fixed price bid, *'...we will bear the burden, and there will be no cost to the government.'*

Nelson somewhat patronisingly complimented Hughes's patriotism, but Hughes retorted brusquely: *'It is not necessarily patriotic. It is my interest in the development of aviation.'*

Nelson again stressed what he saw as the key consideration; *'I think the whole thing is dependent entirely on whether or not a plane of this type made of metal, started now, will be of value in winning a war. Then the other question to be decided is, is it of value to the building of a metal plane to finish this prototype in wood? There is a great difference of opinion on that between you and all of the other people here.'*

Hughes become indignant,*'Well, I just can't help but feel that anyone who expresses that opinion has either a very incomplete knowledge of how the process is gone through in building the first article of any airplane and the testing of it, or else it must be someone who has some reason to see this thing stopped.'*

Nelson asked, *'The question is if in the flying of this wood ship will we gain experience that will enable us to make it better in metal?'*

Late on Friday morning 18 February Donald Nelson called to order the meeting that was intended to have given him the information he needed to decide on the desirability of doing the flying boat in metal and of completing the wood construction as a prototype. The roster of attendees was impressive but did not include Howard Hughes. By the end of the meeting Ridley was convinced by the weight of expert opinion and agreed that there was little value to the wooden construction other than to have finished up what had been started.

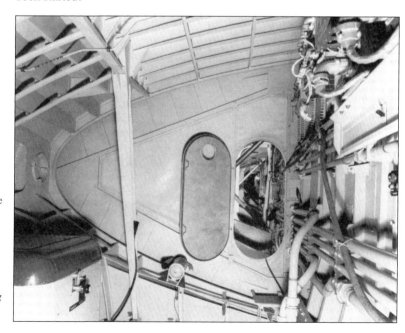

The hatchway into one of the wings. At the point where the wing was attached to the fuselage, the height was 138 inches! *(Aircraft Production - the journal of the aircraft manufacturing industry).*

Another view of the hull under construction in its assembly berth. By now much of the structure is complete, with the solid nose installed, although the flight deck glazing and roof still has to go on. Work in finishing it off is being done on five levels. *(Aircraft Production - the journal of the aircraft manufacturing industry).*

At the conclusion of the meeting Nelson had to rush to a cabinet meeting. Shortly after he got back to his office Hughes telephoned. Hughes said that Ridley told him what happened at the meeting, adding *'...but I didn't know what you decided.'*

'Well, I haven't made a decision on it yet. I am going over to talk to Mr. Jones tomorrow.'

'You call me in the morning any time. I will be here all morning.'

'And will you hold your mind open on it until then?'

'I will hold my mind open on it until then.'

Hughes had been talking with his old friend Jesse Jones, who did not know much about the expert opinion on the flying boat, but was well informed on Hughes's side of the story. After the Friday cabinet meeting, Jones met with the President, to discuss items he did not want to bring up during the meeting. Roosevelt had been interested in the flying boat from the beginning.

Jones's memorandum of this conversation with the President, dictated after he returned to his office from the White House, read: 'I spoke to the President after Cabinet today about the big Kaiser-Hughes plane, and told him that we had been instructed by Mr. Donald Nelson to cancel the contract.

'I explained that the estimate that Mr. Kaiser furnished us at the time the construction of these big planes was being considered was that three of the planes including the prototype would cost $18,000,000, and that approximately $13,000,000 had already been expended on the building, equipment, and prototype; that it was now estimated that to complete the prototype alone would cost probably $5 to $6,000,000; that Mr. Nelson had ordered the contract cancelled due to the extra time it would require to complete it, and upon the advice of experts in the aviation field.

'The President stated that he thought the experience to be obtained by completing one plane would be of too much value to throw away the money already expended, and that the contract should not be cancelled.'

The next day Jones was due to lunch with Nelson. Hughes got in first with a call to Nelson, who had already spent the morning discussing the Kaiser-Hughes flying boat situation with his executive vice-chairman, Charlie Wilson.

Hughes: *'I tell you, I feel very definitely that certain of the so-called experts are prejudiced in this matter, and particularly Mr. Loening, because I would like to call your attention to Loening's report of last August. Mr. Loening's report at that time was most favorable and he said this project should go ahead.*

Now then, in the short period of two months he changed completely, and with no foundation whatsoever that I know of, except the fact that he made a trip up to Martin's and they let him fly the Mars around and rubbed him in the right direction.'

Nelson had to keep his anger in check: *'I can assure you, Mr. Hughes, that is not true. There is no truth to that, honestly. We are not going on Grover Loening's advice alone in this thing. We are going on a whole group, and you are just so dead wrong, Mr. Hughes, really. Honestly, dead wrong, and I can give you my word on that.'*

The conversation went on for four pages of single-spaced transcript and reads as typical Howard Hughes - grinding down an opposition to his throughts. Finally, Nelson's annoyance began to get vocal.

Hughes: *'May I ask you this, Mr. Nelson, so that I can have a clear understanding of this to convey to Mr. Kaiser, because...'*

'Let me talk directly to Mr. Kaiser' Nelson interrupted. *'I have got to go over now and have a luncheon date with Mr. Jesse Jones. I will talk directly to Henry and I will send you a complete transcript of our conversation of last Saturday ....'*

Hughes was forever interpreting conversations to fit his situation: *'My understanding of that conversation was that the going ahead with the metal airplane depended solely upon our submitting a reasonable proposal,'*

'I told you very definitely that the criteria would be whether a metal airplane of the large size could be used in the winning of this war' retorted Nelson.

Complex shapes could be formed using the Duramold process. *(Aircraft Production - the journal of the aircraft manufacturing industry).*

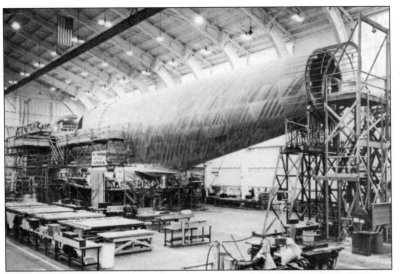

This low view of the rear of the hull clearly shows the sharp 'vee'. *(Aircraft Production - the journal of the aircraft manufacturing industry).*

Hughes continued to talk as if Nelson had said nothing. Eventually an exasperated Nelson finally exploded: *Now Mr. Hughes, honestly, I can't talk any more over the phone about this thing at the moment. I just can't do it. But I honestly think you and I ought to get together instead of constantly talking over the telephone. I don't talk well over the telephone. I like to look at you, like to see you, like to tell you the thing face to face instead of talking interminably over the telephone about it.'*

After two more pages of conversation, Hughes at long last said, *'I will call you then later on.'*

Eventually Howard Hughes finally appeared in Washington to spend five weeks crusading for his flying boat. Nelson called a meeting for 29 February. This gave Hughes an opportunity to more fully present his objections to any cancellation of the HK-1 contract. Present at the meeting were Nelson, Wilson, Burden, Vollmecke, Lewis, Locke, Hughes, Ridley, Loening, Captain Smith of NATS and Colonel Brown of AAF.

Loening, in his notes on the meeting, gave a number of examples of what he called 'the Hughes twist.' *'The general feeling of those present who heard Mr. Hughes, was that he was adroit in twisting arguments in his favor, which did not stand up when analyzed, and unreasonable in his insistence that his estimates were correct, fully to be relied upon and infallible-to such a degree that anyone who questioned them was prejudiced. Also he always had some involved excuse as to why he was late in delivery or had used up so much money.'*

In closing the meeting Nelson stated that he would arrange a meeting the following day with Mr. Hughes, at which Dr. Warner could be present to discuss the features of his report.

On 2 March Nelson wrote to Loening listing a number of points that still troubled him and requested that Harold E. Talbott, who had been Chairman of the 1942 Cargo Plane Committee within the WPB, reconvene the committee to consider promptly the points he had raised.

Two days later Talbott replied that the Committee had met: *'It is the consensus of opinion of the Committee that the construction of the present HK-1 design in metal would not contribute to the War effort. With reference to size of aircraft and general type that would be within feasible range of materially assisting in the War it was the judgment of the members of the Committee that the present work being done for the Army and Navy by the Aircraft Industry, in producing current types and developing new types of cargo aircraft and power plants, is now effective and sufficient for the War needs. It is further noted that since July 1942, a change in the character of the War has, for the present, greatly abated any War need for giant cargo aircraft of spanning transoceanic distance carrying round trip fuel loads. The Committee naturally views with satisfaction the splendid progress now being made on cargo carrying by air.'*

By now it seemed that the Kaiser-Hughes flying boat project had been so thoroughly studied, dissected, analyzed, and worked over that Nelson – who almost certainly had in the back of his mind the activities of The Truman Committee - could not delay decision any longer by asking for another study or another meeting. Clearly, no matter what group, agency, or committee had done the work, published the reports, or held the meetings, the results all favoured cancellation of the project.

What transpired during the first weeks of March 1944 to swing the decision in Hughes's favour? All the facts will never be known. But what is known is that Hughes worked out of his suite at the Carlton Hotel telephoning anyone and everyone who might help him while keeping the pressure on Donald Nelson with that unique Hughes persistence akin to Chinese water torture. And at cabinet meetings Nelson saw the President and Jesse Jones, both of whom were in favour of the flying boat project. Nelson had no desire to cross either the President, Jones, or Hughes; he was definitely caught between a rock and a hard place - it was beginning to look as if it was easier to go along than to resist.

Hughes returned to Culver City about mid-March and fired one last salvo in a letter to Nelson on 17 March: *'I am writing this letter to advise strongly against stopping construction of the HK-1 flying boat now being built by Hughes Tool Company for Kaiser-Hughes.*
Either it was a mistake to start building this airplane or it is a mistake to stop its construction in its present unfinished state .
'If we are going to keep abreast of development in aviation, then we must reconcile ourselves to the necessity of building bigger and bigger

Once the aircraft was sanded smooth, it was then coated with rice-paper before being painted with several coats of aluminised varnish. *(Aircraft Production - the journal of the aircraft manufacturing industry).*

airplanes. This being true, why throw away the $14,000,000 already expended on the HK-1 and later start from scratch on another?

I feel I have made certain contributions to aviation in this country and that you are aware of this fact. I believe I know a good deal about designing and constructing airplanes, probably as much as others who advise scrapping the HK-1 .

I do not believe anyone can say definitely when the war in the Pacific will end. Irrespective of this, our military and naval services are both going ahead with big airplanes which are behind the HK-1 in state of progress.

I sincerely hope you will rescind your order to discontinue construction of the HK-1 and will allow it to be completed on the basis outlined in this letter.

Because of my company's desire to be of every possible assistance in winning the war, it is willing to complete the prototype at the total cost to the government of $18,000,000, provided my company may have the privilege of using the plane at customary rental rates (same to be satisfactory to the government), for a sufficient length of time so that the rental will reimburse the excess cost to my company in completing the plane, such reimbursement not to exceed $2,000,000. Our engineers estimate that to complete the plane will cost more than $2,000,000, in addition to the government's expenditure.

The hull is close to completion, with just work on the flight deck glazing to be finished. *(Aircraft Production - the journal of the aircraft manufacturing industry).*

On 27 March 1944 the contract was reinstated, but with Hughes alone. Kaiser was out. Work continued on the massive components of the flying boat without further serious threats of cancellation. A competition was held among Hughes employees to rename the aircraft. The winning suggestion was Hughes H-4 Hercules, but to Howard it would always be simply the Hughes Flying Boat.

Problems to overcome
Work could now proceed at a pace, but there still were thousands of problems to overcome. As the final assembly, outfittings, and

instrumentation neared completion, Hughes and his engineers found that they would have to redesign the flight control system.

From the beginning, they knew that they had designed an aircraft that was too big for a man to operate the ailerons, rudder, and elevators unaided. What was needed was a powered flight control system that would transmit exactly the smallest movement of the controls by the pilot in the cockpit to the control surfaces, but with a force multiplied two hundred times. Unfortunately, their first design did not respond quickly or accurately enough in the simulator at Culver City.

In April 1946, just two months before the flying boat was moved from Culver City to Terminal Island, David Grant, a Wright Field specialist in hydraulics and controls whom Hughes engineers had previously consulted on the XF-11, was persuaded to work for Hughes on the control problem and was given the title of hydraulic staff engineer. Hughes told Grant that he was to be his co-pilot throughout the flight test programme, partly because he did not want a regular copilot getting in his way and partly because he wanted someone there who knew about what Hughes regarded as the most critical part of the design at that point When Grant started working for Hughes, the control system was really holding everything up.

Grant used some of the technology developed by Kelly Johnson and his team out at Lockheed on their Constellation, but scaled up. The problem also was not helped by the fact that Howard Hughes was a control perfectionist. He wanted more response than he really needed.

The cockpit flight controls themselves were conventional. A column and wheel controlled the elevator and ailerons. Rudder pedals moved parallel to the deck. But it was the manner in which control movements were transmitted to the control surfaces that was innovative. Electrically driven, high pressure hydraulic pumps provided the operating power for telecontrol systems. The pilot's movement of his controls actuated sensitive transmitting relay valves that metered pressurised hydraulic fluid into lines leading to the receiving relay valves at the control surfaces.

These receiving valves in turn admitted pressurised oil to the power cylinders that actually operated the control surfaces.

Although conventional 3/16-inch diameter control cable also directly connected cockpit controls to the control surfaces, they did not by themselves move the surfaces. They only provided a follow-up that ensured the proper relationship between the pilot's control positions and the actual deflections of the surfaces themselves.

For safety, each control surface was operated by two completely independent self-contained telecontrol systems supplied by electric power from separate generators. If one system failed, the other was capable of automatically providing control.

Hughes was also fussy about instrumentation. He spent a great deal of time getting the kinds of indicators he wanted and on their precise placement. Chris Reising was tasked with making sure everything was just so. *'My contacts with Howard were not involved with the electrical system as much as they were with instrumentation and the arrangements for the instrument panel. He was very meticulous in how he wanted the panel arranged. It took long, arduous pains to get the instrumentation just to his liking. He wanted it almost foolproof so that you could follow the flow of the systems.'*

The hydraulic simulator for the HK-1 Hercules. On board, two hydraulic systems powered by the eight engines moved the wing flaps. For every pound of pressure exerted on the control stick, 120 pounds were delivered to the flaps.

The CO2 fire extinguishing system was very elaborate. The hydraulic system instrumentation showed the flow and the switches involved. The air pressure system that charged the hydraulic reservoirs was also very meticulous in design. He wanted those just precise. Hughes went over them in several meetings and marked up drawings and then he'd come back again.

Reising again: *'We had two or three such meetings before he finally approved each system's control and instrumentation layout. There were usually four or five people at these meetings. The drawings would be laid on the table and he would go into the systems in great detail. He had very good basic technical knowledge. He knew what he wanted and belabored each point until he got exactly what he wanted - and he usually had a good sound engineering principle behind the way he wanted it. He would work on us, discuss it and talk about it for hours. He wasn't hard-nosed about it. He wanted to come up with the best design regardless of who suggested it.'*

The Problem With Howard...

When looking at the story of Howard Hughes' business activities, it is clear that the views of others of the man as a manager, entrepeneur and engineer were decidedly polarised. Some saw him as an indecisive, obsessive 'tinkerer' - who worked strange hours and for weeks on end almost impossible to get a hold of. Others as someone who gave considerable freedom to his staff to do what they thought was right.

It has been said that he had Dissociative Identity Disorder, previously known as Multiple Personality Disorder, but it's hard to see if that really was the case. It seems more likely that Hughes preferred to show different facets of his personality, dependant on the circumstances involved and those involved.

The only way for staff to reach Howard was through 'the boys in the back room' at his operations centre at 7000 Romaine Street, Hollywood. The usual procedure was to telephone 'the boys' and wait for a return call from Hughes.

By now 7000 Romaine contained offices for Noah Dietrich, Howard's

Howard Hughes and Odie Odekirk (right) along with another engineer review drawings of the H-4.

financial management wizard; Thomas A. Slack, the Houston lawyer who was one of Howard's principal attorneys; A. V. 'Vic' Leslie, who later became TWA's vice president of finance; Nadine Henley, Howard's personal secretary; Frank William 'Bill' Gay, known by many as 'The Chief Mormon' who handled confidential affairs and was in charge of the back room - 'the boys'. It was Gay who put together the so-called 'Mormon Mafia' that comprised Hughes's inner circle in his later life. They were a small cadre of young Mormons, former factory-hands, potato-chip salesmen and construction workers who Howard Hughes considered especially trustworthy because of their faith and beliefs; Mormons have a health code that eschews alcoholic beverages, tobacco, tea, and other addictive substances. They tend to be very family-oriented, and have strong connections across generations and with extended family, reflective of their belief that families can be sealed together beyond death and throughout eternity. Mormons also have a strict law of chastity, requiring abstention from sexual relations outside of opposite-sex marriage and strict fidelity within marriage. 'The Boys' had been carefully selected whose honesty and sobriety was beyond question and could be trusted with confidential information – aspects that Howard Hughes had grown increasingly concerned about.

Nadine Henley was an interesting character, and for many years was one of Hughes' chief 'protectors'. Born Nov. 8, 1906, in Bloomington, Ind., Miss Henley first worked for a bank and then for Lehman Brothers in Chicago. In 1936, she became one of the first women to trade stock on the stock exchange floor.

Miss Henley began her career with Hughes in 1940, when she moved to California and worked in the engineering department of Hughes Aircraft. She became Hughes' private secretary in 1943, and developed into his trusted aide and confidante.

Thomas A Slack with Howard Hughes at the Senate hearings.

Shortly before the HK-1 project got underway, Howard Hughes brought in an outsider from the automotive industry, John LeDuc, who had no aircraft and no woodworking experience. As works manager, LeDuc would be in charge of the manufacturing organisation. Kenneth Ridley, a young aeronautical engineer who had come to Hughes from Douglas about four years earlier, was placed in charge of the engineering group.

Effectively LeDuc and Ridley held parallel positions within the Hughes organisation. This seemed an impossible arrangement to the supervising engineer of the Defense Plant Corporation's Culver City office. He made an effort to get Neil S. McCarthy, the executive vice president of the relatively ineffective and inactive Kaiser-Hughes organisation who had also been Hughes's personal attorney since about 1925, to bring in a general manager to properly coordinate all efforts.

For a time LeDuc and Ridley appeared to work well together and LeDuc, a man with great drive and energy, began to get the group organised. However, LeDuc had difficulty finding good personnel - after all, Hughes was not supposed to raid the aeronautical industry - and, being

Howard Hughes inspects the flight engineer position on the H-4 Hercules.

thrown into a new and unfamiliar field of work, he made many mistakes. As a result, the engineering and manufacturing departments naturally began to split into two cliques that some refered to as 'the hometown boys' and 'the outsiders.'

In May 1943, Hughes brought in Edward H. Bern as general manager over LeDuc and Ridley. Bern had long experience in American Airlines and had a background in sales and promotions. He moved aggressively to organise Hughes Aircraft and to push the work, but he soon had a falling out with Ridley. But Howard Hughes, because of his longer association with Ridley - and also because of his undoubted penchant for preferring to deal with engineers - began discussing management problems with Ridley without the participation of Bern.

Finally, Bern had enough. On 27 August Bern telephoned WPB Chairman Donald Nelson. *'We have got a terribly chaotic situation out here. It is going to blow right up in your face .... I have been in the business for twenty-six years and I am trying to clean things up and they will not let me. Howard interferes. Howard owns the company. The only time we see him is at home late at night - at ten, eleven, twelve or one or two in the morning. He calls you up and tries to get the picture... The last man that talks to him of his old bunch is the one who can sell him the idea of doing something.'*

Bern resigned that same day, saying that Hughes insisted on organisational changes which he, Bern, felt rightfully fell within his remit as general manager. This was just one resignation in a series that summer. LeDuc, works manager, resigned the same day. Mathewson, head of production engineering had resigned in June. Morgan, night superintendent of manufacturing, was discharged on 18 June. Armstrong, general superintendent of manufacturing, also resigned in June. Blydenburgh was promoted to head production engineering in June. Rivers, general superintendent of manufacturing, was hired in July...

Dick Murrow briefly replaced Babberger as chief of aerodynamics, when Babberger went over to 7000 Romaine to work on the XF-11 design. In later years Murrow went on record attempting to explain the delays and chaos that surrounded the flying boat project: *'I got worried because the project was getting so far behind schedule, and actually it was because Hughes wouldn't relinquish control of things to his people out there. He would spend all night worrying about the most minute details that any young engineer was quite capable of handling. He insisted on digging into details way beyond what he should. We'd spend sessions up there in the middle of the night and then have to be at work at 7 o'clock wartime in the morning, and it was driving us crazy. Then he would disappear for weeks at a time and nothing could get done. I began keeping a set of hand written notes on the meetings-what was decided, when they took place, and so forth-because he got mad at his chief engineers and wouldn't talk to them, first Stan Bell and then Ridley. Us hirelings, like the heads of structures, aerodynamics, powerplants, and whatnot, would have these meetings with Hughes and then we'd have to go in the next day and tell the chief engineer what the hell the next step was. It was kind of ridiculous.*

Hughes was constantly hiring new plant managers, invariably telling them they were in complete charge of things down there. They'd come in

with a bunch of their cronies and set up shop. We used to call them
'torpedoes.' They sat around at their desk with their hat on their head
with a secretary and shuffled papers, and it wasn't long before they found
out that they weren't running the place at all, that Hughes was. And so
he'd get rid of those and bring in another bunch. Odekirk was the real
plant manager for Hughes.

In fairness, it must be said that the complex, often contradictory,
aspects of Hughes's personality, character, and behavior - which caused
such trouble for his managers - included strengths as well. One of
Hughes's senior engineers said, 'I can't give him enough credit. He had
the ability to bring out the best in anyone.'

Hughes worked very closely with his people in matters he was
particularly concerned with, and gave considerable freedom to his
division chiefs. He allowed people he liked and trusted to do what they
thought was right. If he thought that they were onto something, he saw to
it that they had the means to run with it.

When working with his engineers on the flying boat, Hughes did not
worry about the bottom line. Function and performance were what
counted in design, not money. Carl Babberger is on record saying that
the Kaiser-Hughes project had the finest machine shop possible.
*'Anything we needed we got. That's the truth. We were the first people in
Southern California, l dare say, who got the Swiss-made SIP jig borers
that could bore holes to a ten-thousandths of an inch center line spacing.'*

Interestingly though, although Hughes did not skimp on equipment
and materials for the design and construction of aircraft, he was chiseling
and penny-pinching in his business dealings with others, and in the
payment-or rather nonpayment-of taxes.

Hughes often raised hell with his financial people if in purchasing,
accounting, and contracting if they did not extract the maximum possible
financial advantages for Hughes Aircraft. On the other hand, according
to longtime Hughes employees, he paid good wages. People he liked -
people who had done a good job for him - stayed on the payroll even if
there was nothing for them to do.

In general, Hughes treated his working-level employees well,
although he was often inconsiderate in his demands on those closest to
him. Comments by longtime employees are revealing. Jim Dallas,
electrical engineer: *'I came over from Lockheed. Worked for Hughes
twenty-nine years. I always found he was a real fine gentleman and very
understanding. Most of the stuff you hear about him is pure newspaper
bunk. One time I was working at the navigator's station on the flight load
indicators Howard wanted to monitor control forces on the flying boat.
Hughes came by and said, 'What the goddamn hell is all this stuff?' Then
he smiled. I started to explain and then I could see that he knew.*

*He was a friendly guy. In actual design work, he was guiding. He
was the authority. He was right in there all the time. He'd come down
and stay all night, you know. Go over every drawing on the drawing
board. He never went into a meeting or a conference that he didn't spend
a lot of time so he knew everything up and down more than all the rest of
the people there put together. That was actually the secret of his control'.*

Dave Grant, hydraulic staff engineer: *'He really was a perfectionist
and he really was in charge, but he was easy to work for. On the other*

hand, there was the primary difficulty of his weird hours and the fact that if you didn't agree with what he had to say it took quite a few hours of discussion to get to the final conclusion. But he didn't keep us up eighteen hours at a stretch. We were usually through by two in the morning.'

Chris Reising, electrical engineer: *'People would make suggestions and he would review those and throw in his own ideas. If he liked an idea that came up he would so indicate it. He wasn't hard-nosed about it. He wanted to talk about it. He wanted to come up with the best design regardless of who suggested it and he didn't care how long it took. At my level, which was about three steps below reporting directly to the project manager, he was very kind, very thoughtful, very considerate. But if he thought anybody was trying to put anything over on him he indicated his displeasure.'*

Bill Noggle, hydraulics mechanic: *'He was very soft spoken. He wasn't bossy at all. He knew his business, was very much interested in the details, and was very persistent.'*

Homer ('Dave') Roe, powerplant engineer on Hughes Models D-2, F-11, and the HK-1 flying boat: *'I had personal contact with Mr. Hughes on numerous occasions. I found him to be highly motivated and very involved with his projects and felt he was an outstanding leader, He would direct numerous unusual assignments and was technically very competent. He did not engage in small talk, but at times broke through with subtle humor.'*

Reising recalls an episode that illustrates Hughes's relationship with his engineers: *'There were two auxiliary hydraulic pressure systems with separate reservoirs pressurised by the air pressure system. Originally we planned on two air pressure gauges side by side. But in going over the drawings with him on a Sunday morning he asked me if we could get a*

Despite the many tales that Government officials and Air Force representatives were kept out of the Culver City plant, it's clear from this picture that it was not always the case, as Odie Odekirk shows visiting USAAF officials around the flying boat under construction.

single dual air pressure gauge. I said yes.

Because there were so many dual pressure gauges, no problem. So he looks at me and he says, 'Are you sure we can get a dual air pressure gauge?' And I told him yes.

I wasn't about to say no.

During the whole session while we were reviewing that panel he asked me no less than five times point blank, 'Can you get a dual air pressure gauge?' And I said yes every time. He didn't remember ever seeing one.

I couldn't wait to get into work the next day-and there was no such animal! There was just no dual air pressure gauge available anywhere. No one had ever heard of one. So what we got was a dual oil pressure gauge and converted it to the proper pressure and that's what we used. I think that he knew that there was no such thing and was testing me.'

Another of his senior engineers observed that Hughes wanted a day's work for a day's pay. Punctuality meant little to him as long as you delivered a day's work. Although he wanted personal loyalty, he did not like yes-men. He respected those who stood up to him.

'You could argue with him until doomsday. But he was stubborn and persistent and usually got his way in the end. He'd give you hell for failing to do what he told you to do, but not for making honest mistakes.'

Hughes was a perfectionist, extremely fussy about the smallest details; a good example of this was when additonal engines for the flying boat had to be brought by rail from the Naval Air Station in Norfolk, Virginia, Hughes worried that the bumping across the country with the crankshafts in one position would harm the bearing surfaces. So he arranged for them to be shipped in a special car equipped with devices on each engine that slowly rotated their crankshafts during the entire trip and special instrumentation that recorded the speed of the train, the bumps, the starts, the stops, everything, all the way from Norfolk to California.

Howard Hughes in a relaxed mood in his office at the Culver City plant.

When the engines arrived in Long Beach they were torn down completely, inspected, and reassembled. The engineers then were ordered to take them one at a time up to Culver City and run them through a special run-in procedure on the test stand there. Hughes ordered that this could only be done from twelve o'clock at night to eight o'clock in the morning providing the wind was lower than three miles an hour. And he was out there to see that it was done correctly!

Business, Pleasure and the Shady Tree

Following his injuries in the crash of the XF-11, Howard grew a moustache to hide the scar on his upper lip and there were undoubtedly psychological wounds that were never revealed, and which were to haunt him for the rest of his life.

As he grew stronger, he worried that the accident might have made him apprehensive about flying. He had to prove to himself that he

could still fly, not only for his own personal image of the daring aviator who had walked away from crashes, but also to dispel any doubts about his competence as a pilot. On 9 September 1946, photographers snapped pictures of Hughes at the controls of a Douglas B-23 Dragon that had been converted for private use by the Hughes Tool Company. He then took the aircraft up on his first flight since the crash, and during the next few months, he flew everywhere - to New York, Kansas City, Dayton, and Mexico, this last trip being for a vacation with Cary Grant.

Howard Hughes inspects the auxiliary electric power unit at the aft end of the flight deck area. Hydraulic plumbing and electrical wiring runs along the front face of the wing spar. Access to the catwalk inside the leading edge of the port wing is on the right of this picture.

In January 1947, Howard took Grant, who was one of his best friends and sidekick, on a flight in one of the converted B-23s. They flew east to New York, where Hughes had some business to do, and then headed west and south again. The flight plan called for a refueling stop in Amarillo, Texas. On their way back, Grant and Hughes disappeared in the midst of a raging thunderstorm that had grounded aircraft in six states. Across America, newspapers headlined the news that the famed aviator and handsome actor were missing and believed dead.

Obituaries were rushed into type as executives of the Hughes Tool Company met in an emergency session to determine the legal implications of heirless ownership. They were not aware that the two men, trying to avoid the inevitable and ever-present publicity, had intentionally disappeared, changing their flight plan so they could head for Mexico and a much-needed vacation from the press. They wanted it to be a secret so they pulled strings at various places to stop the news from getting out. The two men only became aware of the furore over their assumed deaths after the maid servicing their Guadalajara hotel room became hysterical when she saw the two 'ghosts' napping in their suite!

Howard Hughes always seems to have mixed business with pleasure from a very early age; and with his massive wealth, he could, for the former, indulge in very 'big business', and, for the latter, enjoy the

company of the most attractive, beautiful - and available - women that Hollywood had on offer.

Occasionally, he could take the business-pleasure relationship to a whole new level of sophistication, by intantly requisitioning one of TWA's airliners, even when it was ready for a scheduled service.

Little known even amongst some of his little-known publicised private activities, was the 'Shady Tree Air Service' based at the Hughes plant in Culver City where the flying boat was being built. Normally, an airstrip or runway is cleared of all close by obstructions, but curiously, a clump of eucalyptus trees on the Western edge of the runway seemed to survive the attentions of the bulldozers.

This was because this was the rendezvous for what a few privileged people who were 'in the know' referred to as the Shady Tree Air Service. Howard and a select few of the staff would arrange their rendezvous with their lady friends, who would wait to be picked up at the Shady Tree by the Hughes private air service. The motley fleet was war-surplus B-23 bombers, but Howard's own Boeing 307 Stratoliner was known to make a pit-stop or two at the Shady Tree. Access from the adjacent road was through a gate. Howard, it seems, had the only key.

The most frequent destination was Lake Tahoe, but Las Vegas, Reno, and other cities were on the non-scheduled and very unofficial map.

And it is here that a very minor, but intriguing mystery surfaces. In both the 1956 Ian Fleming's James Bond novel *Diamonds Are Forever* and its 1971 film adaptation there is a character called Michael 'Shady' Tree.

In both the novel and movie, he was the link in the diamond smuggling pipeline who received the diamonds once they arrived in America. In the 1971 film adaptation, the character was re-imagined as a standup comic performing at the Whyte House in Las Vegas. Operating as part of Ernst Stavro Blofeld's diamond smuggling pipeline, he is murdered by a pair of hoodlums on Bloefeld's pay called Mr Wint and Mr Kidd. Shady Tree was performed by Leonard Barr, himself a standup comic.

Hughes Airport, Culver City - half way along it is a dark snudge, which is a small clump of eucalyptus trees alongside a gate that played the part of air terminal for the Shady Tree Air Service. Building 15, where the flying boat was built is on the left.

It is not known if Fleming and Hughes ever met, but Hughes' friend Cubby Broccoli brought the James Bond films to the big screen. Broccoli had met Hughes while working as an assistant on the Hughes 1941 production *'The Outlaw'* starring Jane Russell, and they remained lifelong friends. Intriguingly, the character Willard Whyte was a thinly-disguised adaption of Howard Hughes – was this Fleming and Broccoli 'tipping their hats' to what they knew about Howard Hughes?

Another trait that caused consternation at Hughes Aircraft was that Hughes never went through channels. He often worked with lower echelon people and then had them pass his instructions to their superiors. He would avoid taking time for first-hand contact if he could use an intermediary. This was something that managers - any manager - did not like because that saw it as usurping their hard-won authority in empire-building! It's easy to see how they saw this as being *'..the problem with Howard...'!*

Douglas B-23 Dragon NR44890, seen here in TWA colours. It may well have been the flagship of the Shady Tree Air Service.

Terminal Island And The Move

It was one thing to design and build the world's largest aircraft it was another to find somewhere safe and protected to test it - but when that was a flying boat and the main construction site was nowhere near anywhere suitable, that opened a whole new can of worms!

After careful investigation, Howard Hughes chose Long Beach Harbor's Terminal Island as the final assembly site. It was protected by two major breakwaters, fronted a water area ample for manoeuvring the flying boat, and was conveniently accessible from both land and sea. The site was procured from the Long Beach Harbor Department by a two-year lease with option to renew for an additional three years. Little did the City of Long Beach realise that their eccentric tenant would occupy the site for more than thirty-three years.

Hughes choice of location for the drydocks on Terminal Islands, Long Beach - Pier E, Berth 120 - now part of Pier T, Port of Long Beach.. (Aircraft Production - the journal of the aircraft manufacturing industry).

Here a 290 foot long graving dock was constructed to allow the wings to be attached to the hull that was sitting below ground level, and then, with the dock gates open, allow the craft to be launched.

Terminal Island, located in Los Angeles County, California is between the Port of Los Angeles and Long Beach Harbor. Major parts of the port are on Terminal Island, as well as the Federal Correctional Institution.

The west half of the island is part of the San Pedro area of the city of Los Angeles, while the rest is part of the city of Long Beach. The land area of Terminal Island has been supplemented considerably from its original size.

The drydock for the hull of the flying boat began life as a modest hole at Pier E on Terminal Island, Long Beach. *(Aircraft Production - the journal of the aircraft manufacturing industry).*

For instance, in the late 1920s, Deadman's Island in the main channel of the Port of Los Angeles was dynamited and dredged away, and the resulting rubble was used to add sixty-two acres to the island's southern tip.

The island was originally called Isla Raza de Buena Gente and later Rattlesnake Island. It was renamed Terminal Island in 1891.

In 1927 a civilian flying facility, Allen Field, was established on Terminal Island. The Naval Reserve established a training centre at the field and later took complete control, designating the field Naval Air Base San Pedro - but the base was also known as Reeves Field. In 1941 the Long Beach Naval Station became located adjacent to the airfield. In 1942 the Naval Reserve Training Facility was transferred, and a year later NAB San Pedro's status was downgraded to a Naval Air Station (NAS Terminal

Not one, but three drydocks were constructed, a large one for the hull as seen here and two smaller ones for each of the wing floats *(Aircraft Production - the journal of the aircraft manufacturing industry).*

Island). Reeves Field as a Naval Air Station was disestablished in 1947, although the adjacent Long Beach Naval Station would continue to use Reeves Field as an auxiliary airfield until the late 1990s. Interestingly - and showing just how rumours lead to myths - I have recently seen tales appearing on the web claiming that Howard Hughes flew the Hercules from Reeves Field!

The island was home to about 3,500 first- and second-generation Japanese Americans prior to World War Two in an area known as East San Pedro or Fish Island. Following the attack on Pearl Harbor, all of the adult 'issei' - or first generation - Japanese males on Terminal Island were incarcerated by the FBI on 9 February 1942. Immediately after the signing of Executive Order 9066 on 19 February, the rest of the inhabitants were given 48 hours to evacuate their homes. They were subsequently sent to internment camps, and the entire neighbourhood was razed. The Japanese community on Terminal Island was the first to be evacuated and interned *en masse*.

Over three hundred construction workers took well over a year to build the facility, including putting lock-gates on the three dry-docks then excavating away the surrounding landfill to allow the flying boat to be launched. *(Aircraft Production - the journal of the aircraft manufacturing industry).*

During World War Two, Terminal Island was an important centre for defence industries, especially shipbuilding. It was also, therefore, one of the first places where African Americans tried to effect their integration into defence-related work on the West Coast.

The Port of Los Angeles and the Port of Long Beach were the major landowners on the island, who in turn leased much of their land for container terminals and bulk terminals.

A contract one of Hughes' staff, Charles W. Perelle had negotiated with the Reconstruction Finance Corporation providing for an additional $1,500,000 to move the flying boat to a specially constructed facility where the flying boat was finally assembled and flight tested was still awaiting Hughes's approval. Perelle had started work as a painter's helper at Boeing

Above: The three drydocks under construction, surrounded by the detritus of the building industry!
The trenches cut to allow construction of each dock would eventually be backfilled up to the walls.
Below: The dock for the starboard wing float. In the background is a building that would be used for offices and
equipment storage. Even later, the whole site would be covered in and the flying boat encased.
(both from Aircraft Production - the journal of the aircraft manufacturing industry).

Above: the starboard wing is prepared for the move.

Below: with police escort and along closed roads, the wings move down the Pacific Coast Highway at walking pace on their way to Long Beach and Terminal Island.

Above: the underside of the port wing - both the pictures on this page are thought to have been taken during a lunchbreak stop in Hermosa Beach. The mounting points for the wingtip float can clearly be seen.

Below: The specially shaped supporting structure carrying the starboard wing sits on the moving dollies which in turn straddle the lane divider strips on the highway through Hermosa Beach.

At two miles an hour over closed roads, there were times when an off-road rest-stop was required to let the traffic clear before embarking on the next leg of the move.

Above: Back on the road again on the way to Long Beach.

Below: The starboard wing, followed by the port one, navigates the turn off the Pacific Coast Highway and on to Santa Fé Avenue. Note how the flaps and ailerons have been removed to reduce the width.

Above: Another view of the corner of Pacific Coast Highway and Santa Fé Avenue. The length of the wing is noticable here, as are the number of people who came out to watch the stately progress!

Below: close to journey's end.

and during the 1930s rose to be a major division manager there, then became general manager of Vultee in the early 1940s, where he introduced innovative assembly-line techniques for aircraft manufacture. When Vultee merged with Consolidated in 1943, Perelle became senior vice president for manufacturing. In September 1944, Hughes finally got his man with an offer of $75,000 a year. Perelle was to be vice president and general manager of the Hughes Tool Company, manager of the Hughes Aircraft Company, and a director of Transcontinental and Western Air (TWA).

Hughes was dragging his feet as the contract stipulated that the RFC

Above: Looks like we got us a convoy! The pair of wingtip floats - or 'pontoons' as Hughes Aircraft called them were twenty-three feet four inches long, and twenty feet nine inches tall - and had a thickness of five feet seven inches take the lead on the pair of wings travelling down Seaside Boulevard, with the camera looking east. In the background can be seen the famous Cyclone racer - a twin track wooden roller-coaster - and other 'Long Beach landmarks.

Below: The wingtip floats in their special cradles.

The vertical fin and vert real of the hull seen laying flat on its moving trolley

would choose the pilots for test flights, which was something that Hughes certain did not want. He had every intention of flying the aircraft himself. Perelle had tried for three months to get Hughes to change his mind, but without success.

The RFC was anxious to get the aircraft away from Culver City in order to sell the big wooden plant that housed the hull and other component parts - RFC at that time had the title to both the plant and the flying boat. But as usual, Hughes got his way.

The next problem to be overcome was how to move all the components of the flying boat from Culver City to Terminal Island. Loading the aircraft onto a barge in Ballona Creek - which was the original Kaiser plan and then moving it by sea was considered, but eventually rejected in favour of a overland move, so Hughes's men studied combinations of highways between the two points using hull size as the governing factor. The planning alone is supposed to have taken two years.

The route finally chosen was down the Pacific Coast Highway - several blocks of trees on Rosecrans Avenue that would have to be trimmed and an

Below and Opposite: No, it's not a wingtip of one of the wings you're looking at, but a head-on shot of the vertical fin in its transport frame. In the towns it passed through nothing could get past, but out in what was then countryside between Culver City and Long Beach it was possible for vehicles to squeeze by on either side, as can be seen.

expected 2,300 power cables and telephone wires owned by twenty-one separate utility companies would have to be moved, lifted or re-routed. It was then through Hermosa Beach, before turning off the Pacific Coast Highway onto Santa Fé Avenue. Eventually the caravan would go down Seaside Boulevard before entering Terminal Island via the Navy's pontoon bridge. This was less costly than what would be required using any other combination of highways.

By June 1946, the assembly and dry dock site on Terminal Island adjacent to the eastern boundary of the Long Beach Naval Base had been built to a point where it was ready to receive the flying boat. On 11 June Star House Movers, low bidders for the move at $16,970, placed the giant 160-foot wing sections on moving dollies at a height just sufficient to clear

The wing floats pass oil platforms on their way to Terminal Island.

Almost as big as an airliners wing at the time, the flaps move in convoy.

Like a moth emerging from a chrysalis the hull of the flying boat peeks its nose out of Building 15 at Hughes Airport Culver City on the morning of 14 June 1946.

Moving off of the airfield and on to the highway - At the time this picture was taken the three lengths of timber that was installed to the nose ahead of and up over the flight-deck glazing had not yet been fitted. These show up on later pictures and were put in place to guide away any power or telephone lines that may be been missed by the moving crew.

The hull was a huge item to be moved through the streets of Los Angeles.

parked cars along the twenty-eight mile route. The cost of the move is another of the minor mysteries surrounding the H-4, for depending on who you read the figure varies. The *Los Angeles Times* for instance reported on 12 June that '...*each wing is 191 feet long, 49 feet wide at the root, and 19 feet high. Each wing weighed 34 tons and was moved by Star House Movers Inc., which got the $140,000 moving job.*'

The next day *Los Angeles Times* reported: '*Inching carefully along a close-guarded route from which 2100 individual power and telephone lines had been raised or lowered to provide clearance, the two mammoth wing sections of the Hughes Aircraft Co.'s $20,000,000 flying boat H-4 last night*

Progress was monitored and filmed from an airship hired by Hughes. *(both Aircraft Production - the journal of the aircraft manufacturing industry).*

completed the biggest airplane moving job in history – 28 miles from Culver City to Terminal Island...'

Then the strange caravan inched out of the plant and on to the highways. A battalion of telephone and power company linesmen preceded the convoy, cutting power and telephone lines that could not be moved in advance, then hurriedly spliced them again as the load passed. Alongside them were members of the Parks Service who trimmed any trees and shrubs that restricted its path. Ahead of them, the police of ten cities and towns, together with California State Highway Patrol officers and sheriff's personnel, blocked off cross streets and re-directed traffic. Overhead the blimp Hughes had bought from the Navy to promote *The Outlaw* cruised majestically while crewmen photographed the move!

The second day of the moving job, was completed without mishap and somewhat ahead of schedule. Fifteen Los Angeles motorcycle officers headed by Lt. L. J. Fuller, and an equal force from the California Highway Patrol led by Sgt. Clarence Martin, along with details from half a dozen other cities and towns rode herd on the novel procession, which traveled two miles an hour.

The wings reached the mainland end of the Navy's pontoon bridge spanning the Long Beach channel at 1 p.m. and halted until nightfall, when a high tide raised the structure to near-level with the approaches.

Before the wings could be moved across the bridge to their destination at the graving dock on the east end of Terminal Island, Navy workmen had to remove railings, signs and a post from the floating roadway. Following the wings were the two wing floats held in a huge wood crate. The second day of the moving job, which began Tuesday, and was completed without mishap and somewhat ahead of schedule.

Despite a high level of security attempting to keep onlookers away, the stately progress of the hull through Los Angeles suburbia attracted thousands of gawpers. *(Aircraft Production - the journal of the aircraft manufacturing industry).*

Officers from the California Highway Patrol escorted all the items and closed off streets to traffic.

Power utility and telephone company workers rode on high-lift platforms on the back of trucks to make sure the cables cleared the hull as it passed underneath.

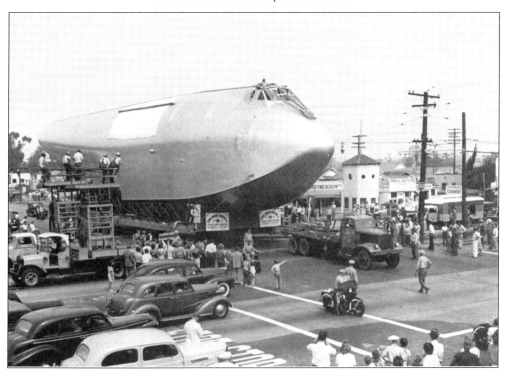

'The flying boat now crossing...'
Above: slow progress down Pacific Coast Highway as utility workers cut cables and re-splice after the hull has passed.

Below: the last hurdle - crossing the pontoon bridge onto Terminal Island.

Toward the end of the route a right angle turn brought the towering hull to a standstill. After an hour of jockeying, of inching this way and that, they negotiated the difficult turn. Plant engineer John Stearns, responsible for the move, felt relief and elation. At no point had it been necessary to remove an additional tree, pole, or other structure, than had been expected and planned for, so carefully had the problem been studied.

Reaching the assembly site at last, they eased the hull down an inclined ramp into its dry dock between the wings laid out in a position for attachment. The long-awaited final assembly was at hand.

Preparations To Fly - Preparations To Investigate

As Hughes readied the giant flying boat for testing in 1947, the special Senate committee to investigate the National Defense Program started to close in and investigate his war-time contracts. It was not unexpected, for three years earlier, Hugh Fulton, then the Chief Counsel for the Committee, sent a letter to E. A. Locke, Jr., the assistant to the chairman of the War Production Board, asking for *'...whatever information is available concerning the status of the Kaiser-Hughes Transport Airplane Project'*. Senators Ralph Owen Brewster of Maine and Homer Ferguson of Michigan, who had been committee members under Harry S Truman, were aware of the cancellation of the contract and its reinstatement in the face of unanimous expert opinion that *'...the contract should be cancelled inasmuch as the project offers no useful contribution to the War effort.'*

On 31 January 1947, Francis D. Flanagan, the assistant chief counsel for the Committee, submitted a memorandum to George Meader, the chief counsel, in which he concluded that: *'...based upon my investigation to date, it is my opinion that this entire flying boat project, which never had the wholehearted approval, and in some cases actually had the disapproval of responsible government agencies, must have received strong backing from some persons with considerable influence in high government places.'*

To Brewster, Ferguson, and all the other Republicans, the Hughes contracts seemed to be low risk, high yield political targets. Not only might they be examples of the kind of wartime profiteering and wrongdoing the committee had been set up to expose, but the investigations could lead to tarnishing the image of Roosevelt's Democratic administration, always a high-priority Republican goal.

Elliott Roosevelt, the President's son, had recommended Hughes's experimental XF-11 as a photographic reconnaissance aircraft for the Air Corps despite the objections of General Echols and other Air Corps officers responsible for the development and procurement of aircraft for the war effort. Also, President Roosevelt interceded in favour of reinstating the flying boat contract after the government's own experts had decided it should be cancelled. Furthermore, although Hughes had received $60 million to produce the two aircraft designs, neither was finished by war's end.

In 1946, Hughes' Trans World Airways had begun flights to Europe in competition with Pan American Airways. Brewster sincerely believed that the United States needed a single American flag-carrier to compete successfully with foreign government-sponsored airlines on international routes. In line with this belief, and in response to special advocacy by Pan American's Juan Trippe, Brewster had sponsored the so-called Chosen Instrument Bill which, if enacted, would have resulted in Pan Am becoming the U.S. flag-carrier in international air transportation.

Vociferous lobbying by Hughes's TWA and other airlines blocked enactment of that bill, but Brewster persisted. Now he was salesman for a somewhat similar bill - often referred to by Howard Hughes as the '*...same baby in a different set of diapers*' that was supposedly Senator McCarran's Community Airline Bill, which if enacted would probably achieve the same end result.

Hughes detested the idea. As the major stockholder in TWA, he had no intention of seeing his airline relegated to domestic service or, even worse, merged with Pan American. However, TWA was in financial difficulties, as Brewster knew well. In January 1947 Brewster, as head of the aviation subcommittee of the Senate Interstate and Foreign Commerce Committee, had taken testimony from Civil Aeronautics Board Chairman Landis in executive session on TWA's financial condition. Brewster knew that the threat of an investigation might persuade Hughes of the merits of the single flag-carrier idea. It did exactly the opposite.

Hughes claimed that in 1945, just after the Civil Aeronautics Board awarded TWA the right to fly the Atlantic, TWA's president Jack Frye had warned him of Pan American's political clout: '*Howard, you are going to learn that Pan American Airways has the biggest, most complex and strongest political machine that has ever hit Washington. Juan Trippe feels you have moved in on his territory, which he considers to be the entire world outside the U.S., and he is going to make your life miserable. You have no idea the lengths to which these people will go.*'

But Hughes could go to great lengths too. His opening gambit in early 1947 was to begin negotiating with Pan Am president Trippe for a merger of the two airlines. Trippe thought he meant it; but Hughes, in devious fashion, was only buying time while laying the groundwork for an offensive of his own.

Meanwhile, in December 1946 Senator Brewster was elected chairman of the Special Committee to Investigate the National Defense Program, as part of the spoils due the Republicans with their new congressional majority. On the morning of 1 February - even before the formal selection of its Democratic membership - Brewster launched into a preliminary hearing to

While Hughes was away in Washington and dealing with some of this other myriad of activities, work continued down at Terminal Island. Here the hull and wings wait for the ramp to be prepared and the cribbing put in place so the hull could be slip down into the graving dock.

The hull is slowly inched into position in preparation for sliding down the ramp into the main drydock

open his promised grand-scale inquiry into the war spending of the Roosevelt and Truman Administrations.

Rumours, leaks, and press speculation convinced Hughes that an inquiry into his multimillion dollar contract to build the wooden flying boat was high on the committee's list of investigations. Hughes flew to the east coast, first to New York and then to Washington. His objective was to forestall a full-scale public inquiry.

The crash of the XF-11

Howard Hughes was juggling a myriad of business interests. ToolCo, movie-making, the H-4 flying boat, Trans World Airlines, and the XF-11 photo-reconnaissance aircraft.

The Hughes XF-11 was a prototype military reconnaissance aircraft, designed and flown by Howard Hughes for the United States Army Air Forces. Although 100 F-11s were ordered in 1943, only two prototypes and a mock-up were completed, with the order for the production aircraft had been cancelled in May 1945

While Hughes had designed its predecessors to be fighter variants, the F-11 was intended to meet the same operational objective as the Republic XF-12 Rainbow. Specifications called for a fast, long range, high-altitude photographic reconnaissance aircraft. A highly modified version of the earlier private-venture Hughes D-2 project, in configuration the aircraft resembled the World War Two Lockheed P-38 Lightning, but was much larger and heavier. It was a tricycle-gear, twin-engine, twin-boom all-metal monoplane with a pressurised central crew nacelle, with a much larger span and much higher aspect ratio than the P-38's wing.

The XF-11 used Pratt & Whitney R-4360-31 28-cylinder radial engines.

Howard Hughes runs up the first XF-11 prior to the first and final fateful flight on 7 July 1946.

Hughes gets airborne from his Culver City airfield. The first aircraft - 44-70155 - was identifiable by the contro-rotating propellers.

Each engine drove a pair of contra-rotating four-bladed, controllable-pitch propellers, which can increase performance and stability, at the cost of increased mechanical complexity. Due to constant problems with the contra-rotating propulsion system, the second prototype had regular four-bladed propellers.

On the urgent recommendation of Colonel Elliott Roosevelt, who led a team surveying several reconnaissance aircraft proposals in September 1943, General Henry 'Hap' Arnold, chief of the U.S. Army Air Forces, ordered 100 F-11s for delivery beginning in 1944. In this, Arnold overrode the strenuous objections of the USAAF Materiel Command, which held that Hughes did not have the industrial capacity or proven track record to deliver on his promises. (Materiel Command did succeed in mandating that the F-11 be made of aluminum, unlike its wooden D-2 predecessor.) Arnold made the decision 'much against my better judgment and the advice of my staff' after consultations with the White House. The order was cancelled on 29 May 1945, but Hughes was allowed to complete and deliver the two prototypes.

The first prototype, tail number 44-70155, piloted by Hughes, crashed on 7 July 1946 while on its maiden flight from the Hughes Aircraft Co. factory airfield at Culver City, California.

Typically for Hughes he did not follow the agreed testing programme and communications protocol, and remained airborne almost twice as long as planned. An hour into the flight - and after on-board recording cameras had run out of film - a leak caused the right-hand propeller controls to lose their effectiveness and the rear propeller subsequently reversed its pitch, disrupting that engine's thrust, which caused the aircraft to yaw hard to the right.

Rather than feathering the propeller, Hughes performed some improvised trouble-shooting - including raising and lowering the landing gear - during which he flew away from his factory runway. Constantly losing altitude, he finally attempted to reach the golf course of the Los Angeles Country Club, but about 300 yards short of the course, the aircraft suddenly lost altitude and clipped three houses in Beverly Hills. The XF-11 was travelling at 155 mph when it clipped the roof of 803 North Linden Drive. The aircraft sliced into the house next door, sheared off a utility pole and finally came to rest, the majority of the wreckage in an alley. The third house was completely destroyed by fire, and Hughes was nearly killed.

Hughes spent thirty-five days in Good Samaritan Hospital. During that time, Odie Odekirk was the only person other than doctors and nurses allowed to see him.

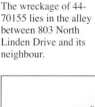

The wreckage of 44-70155 lies in the alley between 803 North Linden Drive and its neighbour.

Guards were posted near the elevators to prevent unauthorised persons from entering sections of the hall leading to Hughes's room, and Odekirk arranged for Marian Miller, later Nadine Henley's secretary, to receive visitors in a small reception room. At times Hughes sent Odekirk out to tell

special visitors that the only reason they could not visit him was that since he was covered with burns and cuts he did not want them to see him in such bad condition. As he recovered he began to fear that his bedside phone was tapped. So he asked Odekirk to make calls for him on an outside phone.

Lying in bed was extremely painful. To move was torture. So Hughes asked Odekirk to have the plant build a new kind of hospital bed with motor-driven segments that would allow him to adjust his body position with no effort on his part other than to finger a control panel. Odekirk relayed his instructions, and Service and Flight Department mechanics built it. The men worked around the clock. Some never went home until the job was done. Some only left for about four hours sleep before rushing back to the job. The bed was delivered to the hospital in record time, but it was never used; it is still in storage in Culver City.'

According to Odekirk, Hughes improved rapidly, but he had an eye on the public impact and drama of his situation. Hughes himself briefed Dr. Vern Mason on what to tell the press and refused to see visitors. Even his mother's sister, Annette Cano Lummis, who had lived with Hughes for a year in California after his mother's death, was turned away. Visitors would diminish the impact of his dramatic fight for life.

The USAAF account into the incident noted that *'It appeared that loss of hydraulic fluid caused failure of the of pitch change mechanism of right rear propeller. Mr. Hughes maintained full power of right engine and reduced that of left engine instead of trying to fly with right propeller windmilling without power. It was Wright Field's understanding that the crash was attributed to pilot error.'*

Dirty Pool and Power Politics.
Much of what follows has been gleaned from the records of the senate hearings, and shows just how Machiavellian and paranoid all sides could be - it was dirty pool and dirty politics that dragged everyone in and down!

Hughes arrived in Washington the weekend of 8 February. He first telephoned Brewster's apartment but found that Brewster was in Kansas

Following his period in hospital and subsequent recovery Hughes successfully flew the second aircraft, 44-70156. This example had only single propellers fitted to each engine.

City. Hughes telephoned Brewster there on Sunday morning to say that he was in Washington and ready to talk to him. Hughes later testified that he desired to make whatever statements were wanted by the committee so that he would not have to return to Washington again.

According to the records, Hughes and Brewster disagreed on what followed this initial contact. Brewster had speaking engagements in Kansas City, Morgantown, West Virginia, and Columbus, Ohio and told Hughes he would not be back in Washington until Wednesday. According to Brewster, Hughes told him, '*I cannot wait until Wednesday. If you will get back here on Monday, I will see that you are flown up there.*' Hughes was to claim that Brewster already held a ticket on TWA to return to Washington on Sunday and accepted free transportation from Hughes by choice, not to accommodate his schedule to that of Hughes.

On Monday 10 February Hughes, accompanied by a local attorney named Hefron, called on Brewster at his office. According to Hughes, he had been advised that Senator Brewster 'was very tricky' and that he should have someone with him to make sure that Brewster could not later claim he had made statements which he did not make.

'*When I met him in his office...*' Hughes was to testify, '*...he immediately*

Looking for all the world like a ship being launched down a slipway - but without the water - the H-4 flying boat hull is slowly slid down into the main graving dock.

Another view of the hull almost in position in the dock on Terminal Island. The two wing floats in their shipping cradles are in the foreground.

launched into the community airline bill. He didn't have much to say about this investigation, he talked about British supremacy, jet propulsion, and labor differential and how it was impossible for the present competitive American systems to succeed because foreign lines could supply services at the lower foreign labor rate, in direct competition with us, and I told him I did not agree.

Furthermore, I did not agree about England's technical supremacy over us. I said that the United States in my opinion always led in commercial aviation and probably always would. I said that some foreign countries exceeded us in flash performance of various military types but that when it came to sound, safe commercial aviation this country had always led, and I felt it always would, and I felt that people might even be willing to pay a little premium to ride on the United States line because I thought it would be safer and the public would appreciate that.

So, anyway, we had some argument on that matter and then I told him I was in Washington and while here I would like to testify before his committee if he so desired and dispose of the matter because I did not want to come back to Washington again. Some time previously I had heard that this investigation was brewing, I also heard that Pan American was behind it, and I thought rather than wait until I was subpoenaed that I would offer voluntarily to testify.'

Brewster testified later that he did not recall the Monday morning conversation quite as Hughes did. Brewster said that Hughes was primarily interested in the investigation. *'I did not know how he had learned about it, but he inquired what it was all about, what we wanted to know, and said that he had come on to discuss this matter with us, and that he wanted a hearing in the matter 'right now.'*

After some discussion Brewster told Hughes, *'I am sure that the committee will be glad to hear your story in executive session. That will not involve any unfortunate publicity for anyone concerned, and I will call the committee together.'*

Hughes arranged for Jack Frye's executive aircraft to fly Brewster to

Each wing, still on its transport dolly, is slowly inched round and into position.

Morgantown that Monday evening for his speaking engagement. The next day the executive session of the committee convened at eleven a.m. to give Hughes his opportunity to answer questions. Hughes hoped that this appearance would suffice. Moreover, he was encouraging Brewster to believe that merger negotiations with]uan Trippe were underway.

Juan Terry Trippe was an American airline entrepreneur and pioneer, and the founder of Pan American World Airways, one of the world's most prominent airlines of the twentieth century.

After graduation from Yale, Trippe began working on Wall Street, but soon became bored. In 1922 he raised money from his old Yale classmates, selling them stock in his new airline, which he called Long Island Airways, an air-taxi service for the rich and powerful. Once again tapping his wealthy friends from Yale, Trippe invested in an airline named Colonial Air Transport, which was awarded a new route and an airmail contract on 7 October 1925. Interested in operating to the Caribbean, Trippe created the Aviation Corporation of the Americas. Based in Florida, the company would evolve into the unofficial US flag carrier, Pan American Airways, commonly known as Pan Am.

Pan Am's first flight took off on 19 October 1927, from Key West, Florida, to Havana, Cuba, in a hired Fairchild FC-2 floatplane being delivered to West Indian Aerial Express in the Dominican Republic. The return flight from Havana to Key West, in a Pan Am Fokker F.VII, took place 29 October, being delayed from the 28th by rain.

Later, Trippe bought the China National Aviation Corporation (CNAC) to provide domestic air service in the Republic of China, and became a partner in Panagra. In the 1930s. Pan Am became the first airline to cross the Pacific Ocean with the famous Clipper flying boats.

Trippe served as the Chairman of the Board of Directors of the airline for all but about two years between the founding of the company and the Second World War. 'Sonny' Whitney, a stockholder, managed to seize this position. He later regretted his action and allowed Trippe to retake it. For a long time Trippe refused to pardon Whitney. At one point, he even agreed to meet Whitney for lunch for a reconciliation but changed his mind and turned around shortly after departing from his office in the Chrysler Building.

Pan Am continued to expand worldwide throughout World War Two. Trippe is responsible for several innovations in the airline world. A firm

Care was needed in positioning each wing structure so that the moving structure missed each dock pit.

The pilot's position of the H-4.

believer in the idea of air travel for all, Trippe is credited as the father of the tourist class in the airline industry, and was the driving force behind Pan Am's formation of the InterContinental hotel group.

The hearing was held in room 457 of the old Senate Office Building. The committee members sat at a U-shaped desk elevated on a dais so they looked down on Hughes, again accompanied by Hefron, who sat at a long table at the mouth of the U-shaped desk.

The hearing proceeded amicably and was concluded that afternoon. At one point Senator Hatch asked: *'Mr. Hughes, you are willing for someone from the committee to come out and go over all of your books and give them all of the information that you have out there?'*

Hughes did not hear the question clearly. *'What is that?'*

Mr. Meader, counsel for the committee, rephrased the question: *'Senator Hatch would like to know if you are willing for a representative of the committee to come and inspect your books concerning this matter.'*

Hughes replied, *'Oh, certainly; of course.'*

There was a misunderstanding here. Brewster interpreted it as an invitation to visit freely and inspect any and all records of Hughes's aviation business. Hughes thought that this invitation was limited to those records pertinent to the specific matter then under discussion. This later became a point of controversy.

After the hearing, Hughes flew Brewster and Senator Bricker of Ohio to Columbus in his B-23 for another Lincoln Club dinner. During the flight Hughes invited Brewster to sit in the copilot's seat. Their conversation was limited by engine noise and Hughes's hearing problem, but according to Brewster there was some mention of the 'community company.'

When the senators returned to the airport following their speeches, Hughes had already departed for Washington and Prye's TWA aircraft was waiting for them. Hughes had decided to stay over Wednesday in Washington and Brewster had invited him to lunch.

Hughes was met at the Mayflower Hotel the next day by both Senator and Mrs. Brewster who soon excused herself and left the men to lunch together in the Senator's suite. As no witnesses were present, the subject of their conversation was known only to themselves - but later one of them was to lie about it under oath. Hughes recalled that he told Brewster he had previously discussed the possibility of TWA merging with Pan American with Juan Trippe during dinner in Trippe's apartment at the Mayflower and that he had promised to give Trippe an answer in thirty days.

Press speculation followed an *Aviation News*, 24 February report: *'The leading proponent of the community airline on Capitol Hill is conducting a one-man missionary campaign on Howard Hughes.*

Senator Brewster and the west coast aviation executive have spent many hours together recently, some of it traveling. Hughes is said to be leaning toward a community company idea and high TWA executives wondered whether it is because of the threat of a full-fledged quiz into the Hughes-Kaiser cargo-plane deal by Brewster's war investigating committee.'

Both men were questioned by reporters concerning the story. In the meantime, Hughes continued his charade of merger negotiations and sent

With the wings now in place, it was time to start to install the tail surfaces.

Juan Terry Trippe (*b.* June 27, 1899 – *d.* April 3, 1981).

Hughes' press agent and general 'fixer' Johnny Meyer.

Noah Dietrich to New York to continue talks with Trippe. Balance sheets were even exchanged. But Trippe's offer of merger was refused as inequitable to TWA stockholders.

On Friday, 14 March 1947 Francis D. Flanagan, former administrative assistant to FBI Chief J. Edgar Hoover and an FBI agent for six years before going to work on the committee in 1944, arrived in Culver City and requested to look at Hughes's records. There was nothing sinister in the timing of this event; Hughes had invited the committee to look at his books. But Hughes later testified that because of its timing in relation to Dietrich's refusal of Trippe's merger offer, he felt the two events were definitely connected.

Flanagan is on record saying that he and Hughes discussed the visit at some length while Hughes was still in Washington and at that time Hughes was affable and cooperative.

Hughes spent a number of hours one day giving Flanagan a tour of the flying boat. His pride was obvious as he tried to impress Flanagan with its virtues. At another point during the two-week visit Flanagan met with Noah Dietrich and a Texas lawyer. Dietrich said, *'Now listen. I know why you're out here, I know what Brewster's after, and that what you're after is Roosevelt.'*

This was news to Flanagan. He and the committee staff knew that Brewster was very close to Pan American people and suspected that Brewster and Hughes might have been talking about Pan American and TWA. But that had nothing to do with the investigation as far as Flanagan was concerned. What he was looking for was how, in the middle of a war, could all that manpower and money be spent on two aircraft designs -neither of which was ever put into production or completed. What kind of mismanagement allowed that to happen? Flanagan even doubted that the original impetus for the investigation came from Brewster.

Dietrich and Hughes made it clear however that they thought that Brewster was using the investigation to pressure Hughes and blacken the Roosevelt administration.

From various accounts, it seems that Hughes's prime motive in inviting Flanagan to Culver City was to head off a public hearing, to talk him around by whatever means into recommending that the investigation be dropped, even if it meant throwing Roosevelt to the wolves. Dietrich is alleged to have said *'You lay off us and we'll give you all the stuff on Elliott Roosevelt. We've got the records'.*

This relates to Howard Hughes' relationship with Johnny Meyer, who Hughes insisted kept detailed, precise expense accounts.

Johnny Meyer, a pudgy, fast-talking, fun-loving exponent of wine, women, and song, had known Hughes for about

eleven years. Meyer's background was in the back-slapping, gift-giving, party-going high-powered public relations world of southern California, first with the Caliente, Mexico Race Track, then as operator of Hollywood's La Congo night club, and then as a public relations expert for two years for Warner Brothers. He then joined Hughes Productions, the motion picture arm of Hughes's empire, on 1 December 1941. After Pearl Harbor Meyer's name was transferred to the Hughes Aircraft payroll, an assignment that gained him the title of 'Assistant to the President and Public Relations Director' which enabled him to obtain six draft deferments during the course of the war.

During the period of their association Hughes and Meyer were often seen together at night spots of the west coast. When Hughes decided to compete on the entertainment front in schmoozing the Air Corps with the other aircraft manufacturers, Johnny Meyer was a natural for the job. It was Meyer who looked after Col. Elliott Roosevelt and his experts from North Africa when they came looking for a photo reconnaissance aircraft. Indeed, it was Meyer who introduced him to actress Faye Emerson, who Roosevelt married four months later.

Installing the vertical and horizontal tail surfaces required considerable scaffolding and heavy lifting equipment.

Assembly Continues

Working conditions out at the Terminal Island site were spartan. The men's room was a five-holer. For the first few months, no shelter at all protected the aircraft. Then a canvas shelter was rigged up on a metal pipe and angle iron frame over the fuselage and each wing.

The exterior finish developed by Hughes consisted of one coat of wood filler, one coat of sealer that acted as a cement for a coat of thin tissue paper placed over it, two coats of spar varnish, and one coat of aluminised spar varnish.

The electrical system devised for the Hughes flying boat was unique: a 120 volt DC three-wire system. It was developed at Culver City by Jim Dallas and L. B. Hillman. They chose a high-voltage system in order to reduce wire size, an important consideration in an aircraft that required very long runs for the wiring. Doubling the voltage reduced the system weight by seventy-five percent.

If a 24 volts system had been used, the main power feeds for the 195 kilowatts required would have had to be solid aluminum rods two inches in diameter and several hundred feet long! So 24 volts was impossible. As to whether to use DC or three phase AC, a cargo aircraft requires ten times more high torque motor applications than AC applications. So a 120-volt DC was used to suit the purpose for which the machines was designed.

Sometimes the engines were called 'corn cobs' because they were huge radials with four banks of cylinders each. On the H-4, this added up to 224 cylinders - or 448 spark plugs and 28,000 horsepower, which was an enormous amount of power. Each engine drove a four-bladed, hydromatic, full-feathering Hamilton Standard propeller seventeen feet two inches in diameter. The four inboard propellers could provide reverse thrust. Propeller spinners incorporated blower fans to help cool the engines during reverse thrust and taxiing. Carburetor air inlet scoops were underslung on each nacelle. The scoop fairings also enclosed the oil coolers. Individual oil tanks in each nacelle were replenished from a fifty-five gallon central oil reservoir by a semi-automatic control system.

All the engine throttles at first were operated by 'pneudynes', devices which used compressed air instead of hydraulic fluid to transmit very small motions a long distance. The receiving pneudyne was supposed to maintain the correct position called for by the transmitting pneudyne despite changes in atmospheric temperature and pressure. The pneudyne system's electrically driven air compressors automatically maintained pressure in the bottled air supply. Moisture absorbing devices protected them against freezing. Since only a small amount of make-up air was required during flight, the bottled air supply would outlast a normal flight even if the compressors failed.

However, the pneudynes did not work as expected. An argument

The Pratt & Whitney R-4360 Wasp Major was a large 28-cylinder supercharged air-cooled four-row radial piston aircraft engine as fitted to the H-4.

developed between project manager Bill Berry and the lab people who had to endorse or approve the system to be used. The key issue was the throttles; the pneudynes just would not come up in the same spot.

Engine mechanic Al Geverink said that once they got the power plants installed and were operating the engines and their associated systems and controls, Howard Hughes came down to the Terminal Island site routinely.

'He was up on the flight deck and we were back of the engines. We had all air controls at that time before we changed over to electricity. They were a pain in the A. It was pretty hard to control them. No valve will operate the same with the same amount of given pressure. The pilot only having four throttles - one throttle operating two engines - why, we had to get back there and adjust them. The flight engineer had eight throttles, but a lot of times we could even them up out in the wing better than he could at the instrument panel.

The pneudyne-operated throttles were used for the one and only flight of the flying boat but were subsequently replaced by electrically operated throttles for Hughes wanted a fail-safe design, so they developed a zoned electrical system with complete redundancy throughout. All electrical subsystems had fallback arrangements. Bill Berry, the programme manager, made a trip to New York City to study how Consolidated Edison handled short circuits while still feeding power to unaffected areas. The resultant design was a three-wire system with current limiters for each wire. If one line shorted, two remained to carry the load. They also were very careful to ensure that all electrical relays would work at high altitudes. Such electrical equipment had to be built especially for this under contract to Hughes by

The overhead views provided regular updates as to the progress of assembly of the flying boat.

Bendix. Backup or standby electrical power was provided by two 30-kilowatt auxiliary power units by Jack and Heintz. In addition, emergency battery power was provided by ten 12-volt batteries in two banks. The electrical servo throttle control system operated from the 115V DC electrical system and consisted of an electrical potentiometer mechanically connected to each throttle lever. The electrical potentiometers were wired to servo units located in each nacelle and the servo units were mechanically connected to the carburettors. Full travel of throttle operation from full open to full close was obtained in approximately one second. When the throttle lever moved the electrical potentiometer, the servo unit positioned the throttle valves in the carburettors which responded to throttle lever motion in .03 seconds. A throttle inching control system operated through an independent system which primarily consisted of a separate motor mounted on each servo unit and controlled from either the pilot's or fight engineer's station by toggle switches. There was a mechanical override provided in each nacelle in case both the primary throttle control and throttle inching systems failed.

As the electricians hooked up the electrical circuits and made functional checks of the electrical equipment, engine man Harry Kaiser and his crew of power plant mechanics put the power plants together, hung the engines, and mounted the props.

The first set of engines was obtained from the Navy on the West Coast. *'We got hold of nine 4360s, the model A, from the Navy in San Francisco and brought them up to the B class. In addition to other modifications, they added different manifolds'* recalled Harry Kaiser, one of Hughes' longtime employees. It was Kaiser who designed and built the engine test stand used

The mid interior, with the lower hull filled with the supposed 'beach-balls' to aid floatation.

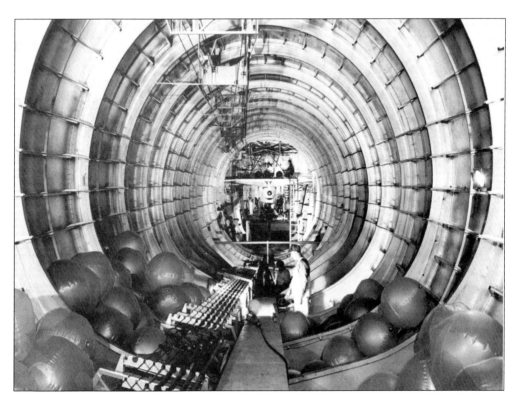

to test and run in the engines used on the flying boat. Later, further sets were obtained from the Navy in Nortfolk, Virginia and shipped by rail to California.

The eight air-cooled, radial R-4360 Pratt and Whitney Wasp Major engines rated at more than 3,000 horsepower each, were accessible for in-flight inspection and minor repairs through a passage in front of the main spar, and then through the nacelles and firewall doors. The wing was thirteen feet thick at the hull. All fuel, oil, hydraulic and pneumatic plumbing was routed along the spar, which provided for quick inspection. All cowling and other structures forward of the firewalls were of metal.

The aircraft taking shape was indeed something special. Its cargo space was equal to that of two railroad box cars. Originally, the design called for clamshell nose doors and a nose loading ramp. These were not installed on the test airplane. The hull's inch-thick bottom skin could take a much higher bottom pressure than that of smaller flying boats. Below the cargo deck the hull was divided into eighteen watertight compartments. If twelve of these were flooded the ship would still float.

As additional protection in case of flooding, Hughes at first had beach balls placed in the hull. Like the ping-pong balls used in his round-the-world Lockheed, they were to provide buoyancy in case of an accident on the water.

Interior photographs of the hull taken during this period clearly show the balls, but the captions sometimes erred as to their purpose: they were not to protect people who fell from the catwalk, as one caption explained. A net placed over the balls ensured that buoyancy would be applied to the lowest points at which the net was fastened to the hull. Prior to the test flight, however, the beach balls were replaced by styrofoam meticulously contoured to the ship and placed alongside the catwalk.

Covers appeared over the wing centre area and flight deck as work went in protected from the weather.

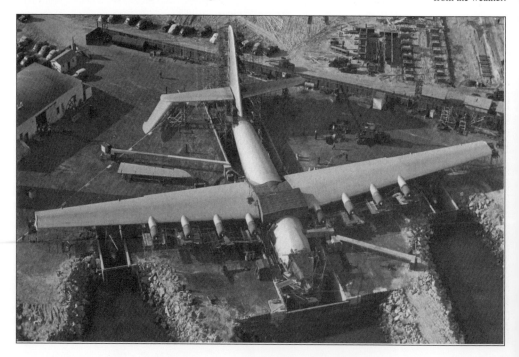

The area around the flying boat on Terminal Island slowly evolved as more and more buildings were erected and fencing was put up landside to protect the boat from prying eyes.

The pits dug for the docks were fitted with lock-gates and then opened up to the harbour in preparation for the flying boats launching.

More Politics, More Investigations

Hughes was convinced he was being pressured by Trippe and Brewster, so his next ploy was to telephone Juan Trippe in New York. *'Look, there is no chance of us getting together unless we discuss this matter in person.'*

Trippe agreed and arranged to meet with Hughes in California. The meeting was held in Palm Springs on 12/13 April in a house in the desert rented by Hughes for the occasion. Trippe arrived from New York in his personal aircraft, Dietrich from Houston in his, and Hughes in his B-23. It seems that Howard was cordial to Trippe, even though he disliked him intensely for his efforts to monopolise overseas air travel for Pan American. The talks continued into the evening of the first day and were resumed on the morning of the next day.

During the meeting, Hughes brought up the question of Brewster's investigation as though this were a matter that Trippe could control. By Hughes's account, Trippe was not bothered by this and talked as though Brewster worked for him. Not long after the Palm Springs meetings Trippe finally realised that Hughes had no intention of merging TWA with Pan American. Meanwhile, the investigation was moving ahead at an accelerated pace, and Hughes realised that revelations of Meyer's expense account would figure prominently.

Hughes was acutely aware that little could be said in defence of Johnny Meyer's activities except that everyone was doing it. Also, he knew that he was vulnerable to critical inquiry into the poor management and delays in fulfilling his two contracts. Clearly Hughes realised that the best form of defence would be a stinging counterattack!

After the first successful test flight of the second XF-11, Hughes spent more time with the flying boat - just at the time when Senator Brewster started leaking charges and salacious items to the press.

The view inside the hull, looking towards the rear from the mid-section. The ladder on the left goes up to the flight deck area, and can also be raised through the top of the hull up to an extendable manoeuvring platform used to get a good overall view for entering or leaving the dock.

The Senate committee to investigate the National Defense Program through Francis Flanagan had obtained most of Hughes' records for their investigations - of course Senator Brewster as Chairman of the Committee knew this. Long after the event, Flancis Flanagan went on record about some of the behind the scenes dealings. *'We never intended to put ninety percent of these expense account items about actresses and whatnot in the record. But we had the originals and made one set of copies which we punched and put in a loose-leaf notebook. One day a member of Brewster's staff asked to see the expense account records, so we sent him the copies with the holes in them. The next day when photocopies of Meyer's expense accounts appeared*

A crane lifts temporary covers over the aircraft to protect it and the workers from the effects of the harsh sunshine.

in the newspapers, sure enough the holes were there'

Brewster called an executive meeting to find out how the leak occurred. and it seems that Flanagan told him the copies printed in the paper were the ones with the holes that his staff had borrowed. *'Well, call off the meeting'* retorted Brewster. *'He knew we knew. He was very brash that way. You'd catch him in something and it wouldn't faze him. He'd just laugh it off.*

The hull resting on blocks inside the main dock.

Hughes remained silent while the scandalous revelations made headlines. Behind the scenes he was using his wealth, connections, and cunning to lay the groundwork for his coming offensive.

Hughes called Hugh Fulton, the former chief counsel for the Investigating Committee who had been replaced by George Meader. Hughes told Fulton that when he had appeared in Washington he had not believed that he needed counsel, because he had thought that the committee was concerned only with the engineering aspects of his aircraft contract with the government, and that he had felt perfectly competent to handle any inquiries along that line himself. Hughes was confident that that kind of an investigation would disclose creditable performance on his part. However, after Flanagan contacted him in Los Angeles and requested Johnny Meyer's expense account, he was convinced that the Committee was out to smear him.'

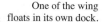
One of the wing floats in its own dock.

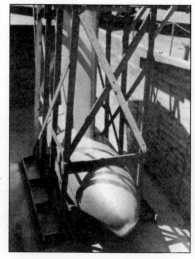

Fulton immediately reported to Brewster: *'It was Hughes's contention that the Committee's investigation of the aircraft contracts was being utilized on behalf of Pan American Airways and the proponents of the Chosen Instrument Bill; that Hughes was aware of a close association between Senator Brewster and Pan American*

Airways, including entertainment of Senator Brewster by Pan American Airways without charge to Senator Brewster; and that in the event of a public hearing which exposed the relationship between Howard Hughes, Brigadier General Elliott Roosevelt, and certain entertainment of General Roosevelt by Hughes and representatives of the Hughes Tool Company, it would be Hughes's intention to counterattack by exposing relationships between Senator Brewster and Pan American Airways.'

Brewster was incensed at the suggestion and stated that the investigation was not one which he originated, that it had been developed by Meader and Flanagan and the staff before he was acquainted with it. Brewster claimed, the investigation had been under consideration by the Committee in one form or another for several years, even back to the time when Fulton was chief counsel. The suggestion that the investigation was related in any way to his advocacy of the Chosen Instrument Bill was incorrect and unwarranted, he exploded, and his attitude toward the progress of the investigation would not be in any way influenced by the fact that such a suggestion had been made.

Fulton duly reported the conversation to Hughes by telegram.

On Easter Sunday, Hughes called Fulton again and asked that he represent him in connection with the Committee's investigation. Hughes told Fulton that he was convinced that the Committee was out to smear him.

On 8 April Fulton sent Rudolf Halley, his law partner and the former associate counsel and chief counsel to the Investigating Committee, to Los Angeles. Halley seemed to think the case was pretty bad for Hughes.

Fulton told Meader that if the Committee seemed determined to go into a full and complete public examination of Hughes's affairs - including transportation and entertainment furnished to public officials - he would be

Howard Hughes standing by the engineers panel.

Once the engines were hung, the propellers could be installed.

Hughes was so protective of the aircraft, he insisted that everyone wore overshoes when they were inside the flying boat for whatever reason.

inclined to represent Hughes. He said further that he was inclined to agree with Hughes's strategy of defending himself by counterattacking on the grounds that this was a political investigation to smear the Democrats, Hughes, and Elliott Roosevelt to the detriment of TWA and to the advantage of Pan American Airways, and to attack Senator Brewster's association with Pan American Airways.

Meader wrote in his memorandum of the event that *'Fulton stated that his only reason for discussing the matter with me was to apprise me fully of all the circumstances and to caution me that in the event of a hot public controversy, I should bear in mind that I might be in the thick of it with brickbats flying.'*

Meader considered that Hughes had brought on what publicity had already attended the investigation by rushing to Washington and insisting on speaking to the senators themselves. This resulted in an executive hearing on which there was some publicity. Hughes's attempt to control the way the Committee conducted the investigation by stating the terms on which his files would be made available to Flanagan resulted in the Committee's decision to issue a subpoena. This subpoena was announced to the press by Senator Brewster.

Meader thought Hughes had been ill-advised in these actions, *'...since the staff had simply been conducting a preliminary investigation, as we have done on a great number of matters which have never been mentioned publicly and concerning which we have not been in the practice of releasing any information to the public until the preliminary investigation was substantially completed.'*

The propellers were just over seventeen feet from tip to tip, and yet look small in this picture.

With Futon and Halley on board, Hughes then set about hiring investigators to dig up evidence of Brewster's close connections with Pan American and of his acceptance of so-called 'gifts and favour'. In addition, he sent Johnny Meyer on an extended trip around South America and to

Europe to uncover evidence of Pan American largess to the Brewsters during their trips abroad. Finally, he contacted journalists and publishers who might help him in his fight.

Even with the vertical fin in position, it still took a lot of scaffolding to hang the rudder.

One important ally was Drew Pearson, a Washington gadfly and muckraking columnist whose *'Washington Merry-Go-Round'* was widely read - and feared. Pearson had often spoke about Pan American Airways with great hostility for wielding such political influence that it was supported by taxpayers' subsidies and protected by exclusive government franchises. Pearson also bore a particular ill-will toward Brewster, whom he often

One of the wing floats or 'pontoons' in its dry dock.

referred to as 'the kept Senator of Pan American Airways.'

Pearson ingeniously wove a large amount of anti-Brewster material into his columns, ostensibly praising Brewster for his work but at the same time adding that it stopped short of investigating Pan American.

Another important ally of Howard Hughes was William Randolph Hearst. Right after the public hearings into Hughes's two contracts opened in Washington on 28 July, Hearst's *Los Angeles Examiner* published a open letter from Hughes to Brewster that was carried in Hearst papers throughout the country and reprinted by other major papers. After days of public silence while the press ran sensational stories about his lobbying activities and Brewster's version of his role in the unfolding melodrama, Hughes was at last striking back. Everything now started to fall into place.

'You know, Senator, if you hadn't gone too far overboard, I might have been willing to take a certain shellacking in this publicity spree of yours.

Yes, I might have been willing to sit back and take a certain amount of abuse simply because I am only a private citizen while you are a Senator with all sorts of rights and powers to subpoena me, make me drop my work, and travel 2,500 miles back there to Washington and otherwise cause me no end of inconvenience.

'So I sat back and let you do the talking for five days.'

The letter in the Hearst newspapers continued in this vein for some paragraphs, making the points that Hughes was not a politician but an ordinary citizen, and that he had been in the airplane business for thirteen years and lost $14,000,000 - *'So I don't think I have been defrauding the Government'* - then the tone moved from defence to attack.

'Why not tell the whole truth? Why not tell when this investigation was really born? Why not tell that this investigation was really born on the day

that TWA first flew the Atlantic? On the day when TWA first invaded Juan Trippe's territory - on the day when TWA first challenged the generally accepted theory that only Juan Trippe's great Pan American Airways had the sacred right to fly the Atlantic?

And why not tell about my answer: That I would have to think it over for thirty days, and that I would let Trippe know at the end of that time?

And why not tell about Mr. Trippe's flight to the Coast to see me, and how Mr. Trippe and Mr. Dietrich and I spent two days at Palm Springs, Calif. and why not tell what we were talking about for those two days?

'And while you are at it, Senator, why not tell about the two airplane trips you bummed off of me?'

With the removal of the earth in front of the drydocks completed, the area was flooded and the H-4 could float free into Long Beach Harbor.

On 30 July Hughes issued a statement to the press that directed four questions at Senator Brewster and demanded an answer:

1. At lunch in your suite at the Mayflower Hotel, Washington, D.C. did you or did you not offer to call off this entire investigation if I would agree to merge TWA with Pan American Airways and support your community airline bill?

2. Did I or did I not reply that I had already promised Juan Trippe that I would give him my answer in thirty days?

3. And did Juan Trippe subsequently make a flight across the country to see me, and spend two days closeted in secret conference with me at Palm Springs, Calif., or did he not?

4. And when you returned from Europe thereafter, did Juan Trippe tell you to hold up and not push this investigation for a while, as he and I were still trying to get together, or did he not?

Each day there were new press statements as Hughes delayed appearing in Washington in defiance of requests to do so by Senator Ferguson. He had taken over as chairman of the subcommittee set up at the request of Senator Taft to handle the investigation so that Brewster would be free to answer Hughes's charges as a private citizen.

'It is a sad situation when a United States Senator has to drag a lot of innocent girls into a Congressional hearing in order to achieve personal publicity' Hughes stated the next day, seemingly prompted by the subpoenas issued for Judy Cook, a swimmer, and Martha Goldwaite, a model.

Hughes also attacked committee demands that he rush to Washington: *'So far I have not had more than two hours sleep for the last week which as you know was one of the hottest in history. In addition, I have not completely recovered from my accident of last year. So it was under these circumstances that Senator Ferguson called me on the phone yesterday afternoon at 5:30 p.m. and told me to appear in Washington before his committee this morning at 10 a.m. I simply replied that under the circumstances I refused to jump through a hoop like a trained seal and fly all night and appear the next morning without any sleep to testify all day on a most important matter. I am a little tired of being pushed around and intimidated by Senators Brewster and Ferguson just because they have some very strong powers which are granted to all Senators but which were not intended to be misused for the promotion of a three-ring publicity circus.'*

Ferguson's response to Hughes's foot-dragging and newspaper salvos from the sidelines was to issue a subpoena for his appearance to be delivered

A general overall view of the Terminal Island site as things come together ready for the first flight. Off to the right floating dock staging for some of the Hughes marine fleet. To the extreme right is the final revelation as to what the cylindrical structure erected on the site was for - it was a control tower!

to Hughes in California the following day, Friday 1 August. Hughes, however, held to the original date set for his appearance. In Los Angeles he stated that he would be in Washington the next Wednesday to answer questions, and that pressure of business would prevent his going there until then, no matter what.

Drew Pearson advised Hughes that he could not win through lawyers, columnists, and PR men; neither could he ride out the storm by lying low. Although the basic ingredients of public sympathy lay on his side, he would have to face his accusers personally.

Hughes agreed, and it seemed that his own experience as a moviemaker prompted him to cultivate the persona and mystique of the western hero who, armed with moral right, rides alone into a town controlled by his enemies for a confrontation at high noon. Hughes slyly cultivated the underdog image: *'Juan Trippe and Brewster made it very clear to me that they only needed TWA in order to put their community company airline bill through Congress and make it a law. They could shove it down the throats of the other overseas airlines if they could just get TWA on their team. Then Juan Trippe, who already has the biggest airline in the world, would wind up with a complete monopoly, automatically and quite legally, putting all the smaller airlines out of business and taking over their routes and airplanes.'*

As US Marshals sought to deliver Ferguson's subpoena into Hughes's hands, Hughes disappeared from public view. Everyday Americans were tickled by Hughes's defiance and by the committee's seeming inability to find him; public interest in his every word and action increased. Moreover, the committee, having to put on a show every day, would soon run through the lesser witnesses, thereby clearing the stage for the climactic event: Hughes's face-to-face confrontation with Senator Brewster.

A confident Brewster in a 2 August radio interview taunted Hughes, saying that he hoped to expose *'...the whites of his lies.'*

Newsmen kept watch on Hughes's B-23 at Burbank Airport hoping to spot Hughes when he left for Washington, so Hughes called Odekirk, and told him to fly the B-23 to Culver City. As the plane climbed out of Burbank, Odekirk banked to the east as a diversionary manoeuvre. He later swung around to drop down unobtrusively and land on the Hughes Aircraft Company's long private runway.

On Monday night, 4 August Hughes materialised briefly in Las Vegas and spent several hours in the public rooms of the Last Frontier. The next day at 2:32 p.m. Pacific Standard Time, he, flight engineer Earl Martyn, and his two bodyguards took off from Culver City for the long flight to Washington. US Marshal Robert Clark grudgingly admitted that the subpoena was still undelivered.

'He ducked' he said.

The Hearings

It was not a quiet flight to Washington. Lightning flashed in a line of thunderstorms ahead. Hughes, who had rested in the pilot's seat while Earl Martyn maintained course and altitude, tightened his seat belt, took the controls, and peered into the blackness. Soon they passed into better weather, but before the night was over there would be more.

The whole saga was recorded in 'Investigation of the National Defense Program - Additional Report[s] of the Special Committee Investigating the National Defense Program, Pursuant to S. Res. 71 (77th Congress, S. Res. 6, 78th Congress, S. Res. 55, 79th Congress, and S. Res. 46, 80th Congress) Resolutions Authorizing and Directing An Investigation of the National Defense Program. Pt. 40 Aircraft Contracts (Hughes Aircraft Co. and Kaiser-Hughes Corp) which has been reported here verbatim in italics.

The First Day – Monday 28 July 1947.

By daylight the storms were behind them. The weather was clear as they descended across the Georgetown Homer, a holding fix for aircraft waiting their turn for instrument approaches to Washington National Airport during bad weather. Washington Tower cleared them for an immediate visual approach to runway 18.

Hughes lowered the landing gear as they crossed over the Memorial Bridge not yet crowded by rush-hour traffic, made a shallow right turn to the runway heading as they passed over the 14th Street Bridge, followed by his usual smooth touchdown. Turning off the runway, he taxied immediately to the TWA parking ramp.

After braking to a stop and shutting down the engines, he jackknifed his tall frame through the aft door of the B-23 and half-jumped to the ground. He was wearing his usual hat, brim snapped down in front, a white dress shirt open at the neck, and a grey double-breasted suit with coat unbuttoned. Grim faced, he strode past the cameras.

Noah Dietrich, who had met the aircraft, manoeuvred Hughes into a waiting car, and they headed for the Carlton Hotel. Hughes had promised to appear before the committee at ten that morning, but he was exhausted. *'Noah, I've got to get some rest. Ask them to postpone the hearing until afternoon'.*

In the city, the Subcommittee of the Spectial Committee to Investigate the National Defense Program gathered. A transcript survives that records the occasion.

'The subcommittee met at 10:30 a. m. in the caucus room, Senate Office Building, Senator Homer Ferguson (chairman of the subcommittee) presiding.

Present: Senators Homer Ferguson, George W. Malone, Joseph R.McCarthy, Carl A. Hatch, Herbert R. O'Conor, Owen Brewster, Claude Pepper, Harry P. Cain.

Present also: William P. Rogers, chief counsel to the committee. Francis D. Flanagan, executive assistant to the chief counsel of the committee.

Senator Ferguson: *The committee will come to order.*

At the opening I think it would be well to announce the purposes of these hearings. The purpose of these public hearings is to make inquiry into certain contracts between the Federal Government and the Kaiser-Hughes Corp. and the Hughes Tool Co.

The first phase of the hearings will be devoted to the contracts for the construction of flying boats. The committee has been advised that in November 1942, the Defense Plants Corporation entered into a contract with the Kaiser-Hughes Corp. for the construction of three large flying boats which were to be built on a nonprofit basis at a total cost not to exceed $18,000,000 It was contemplated that the first of these flying boats would be ready for flight in 21 months.

In March 1944, after more than $13,000,000 of government funds had been expended on the project and the first plane was not yet completed, the Kaiser-Hughes Corp. withdrew from the project. At that time a new agreement was entered into between the Hughes Tool Co. and the Defense Plants Corporation. Under this new management, the Hughes Tool Co. was to complete the one flying boat then under construction at a cost not to exceed $18,000,000, including the $13,000,000 already expended.

This flying boat project was initiated to aid in the war effort. Now, 2 years after the end of the war, and almost 5 years since the beginning of the project, the boat has not yet been flown. The Government has already expended over $18,000,000, and is obligated to expend additional funds for testing the plane.

The subcommittee will also examine the Procurement of the Hughes photo-reconnaissance plane known as the F-11.

The War Department entered into a contract with the Hughes Tool Co. calling for the construction of 101 photo-reconnaissance planes, at a total cost of about $70,000,000. Ninety-eight of these planes were to be production models, one was a static test model, and two were experimental planes. Although the definitive contract was not finally approved by the War Detpartment until August 1, 1944, the Hughes Tool Co. began work on one photo-reconnaissance plane under a letter of intent dated October 11, 1043. At that time plane deliveries were scheduled to begin within 12 months.

In May 1945, the War Department canceled that part of the contract covering the 98 production models, but the Hughes Tool Co. continued the construction of two experimental planes. On July 7, 1946, the first experimental plane crashed on its initial test flight. The Hughes Tool Co. is now testing the second experimental plane, but that plane has not as yet been accepted or delivered to the Army Air Forces. Approximately $22,000,000 of Government funds have been expended or obligated on this photo-reconnaissance plane project. This amount includes $8,600,000 in termination costs.

It is the intention of this subcommittee to thoroughly examine these two wartime procurement projects involving some $40,000,000 of public funds. We intend to call a number of witnesses for the purpose of taking testimony regarding all phases of these contracts, including pre-contract negotiations, the award, performance, and termination of the contracts.

We have here this morning Mr. Meigs, who was connected with the War Production Board at the time that the first contract was made.

During these hearings we will probably be referring to these contracts under the names of the 'Flying Boat contract' and the 'F -11 contract,' which were the names usually used to describe these two planes.

Mr. Meigs, will you take the witness chair here, please! If you have anyone with you that you want to get information from, Mr. Meigs, you can have them sit at the table with you.

Mr. Meigs. I do not have anybody, sir.

Hughes's attorney, Thomas A. Slack of Houston, asked to be heard. *'I wanted to tell this committee, that pursuant to its request, Mr. Hughes is here and ready and anxious to testify before the committee whenever it suits the convenience of the committee.'*

Ferguson observed that 'here' must mean in Washington 'rather than present in this room.'

Slack was ready for this sort of comment. *'Last night we read in the press that you had issued the statement that Mr. Hughes would not be used this morning, that he would have to await the finishing of testimony of other witnesses, and may I say that it was on my own suggestion that I make this announcement to you, and that Mr. Hughes is awaiting your pleasure and he will be available on your request at any time.'*

This left Ferguson little option but to say that *'...rather than delay until he gets here we will proceed with this morning's proceedings . . . and then recess for an early afternoon session at which time Mr. Hughes will be called as the first witness.'*

This gave Hughes an advantage. Rather than having to wait for Johnny Meyer and Roy H. Sherwood, the assistant comptroller of the Hughes Aircraft Company, to finish their testimony, he could enter as the star attraction for whom everyone was waiting. He derived further psychological advantage by arriving forty-two minutes late for the afternoon session, which had been scheduled for two o'clock.

At 2:15 p.m., Senator Ferguson banged the gavel and called the committee to order. *'I want to announce that Mr. Hughes will be here at two-thirty when the session will start, and I will put on the record now a memorandum that has been handed me by the War Department ....'*

With that business concluded, he again recessed the Committee to await the arrival of Mr. Hughes.

While the members of the Committee idled at their places, the waiting spectators, who were jammed into the high-ceilinged, chandelier-lit caucus room, eyed the great oak entrance doors expectantly. As the time ticked on, senatorial majesty wilted; but for the spectators the approaching moment of Hughes's entrance became even more dramatic.

As two-thirty came and passed, Ferguson was visibly restive. *'Mr. Slack, what is the latest news from the front?'*

'I reported to you the latest news, Senator, that they were leaving right away."

The car carrying Hughes and Dietrich from the Carlton Hotel proceeded up Constitution Avenue toward Capitol Hill, swung left at Delaware Avenue, and stopped at the corner. Newsreel cameramen followed every move.

Hughes got out, tucked a small stack of folders and papers under his arm, and walked toward the building. A press photographer stepped into his view as Hughes and his small entourage passed a group of women waiting on the sidewalk. A flashbulb flashed.

The group proceeded directly across the walk, up the marble steps, and through the main entrance at the southwest corner of the Senate Office Building into the domed rotunda. The route was lined with well-wishers and the curious.

As he approached the massive oak doors of the ornate second floor caucus room, he could see that he was playing to a standing-room-only crowd that trailed out into the corridor and snaked out of sight into the corridors beyond.

Loudspeakers had been set up so that those standing in the corridors could follow the testimony given inside the room. As Hughes entered, the crowd clapped and welcomed him noisily. It was, said *Newsweek,* the biggest circus that had pitched its tent in Washington in many years. Six movie cameras and one television camera whirred. It was the first congressional hearing to be televised and was being watched by five thousand pioneer viewers in Washington and New York.

Senator Ferguson pounded a glass ashtray on the desk. The spectators watched the lanky, dark-mustached man in a loosely fitting grey suit, his neck sticking out of a too-large collar, as he made his way through the room. Twenty press photographers rushed to surround the witness table. Hughes deposited his stack of folders on the table before him.

Senator Ferguson spoke: *Let it be known that we must have quiet in the room. Mr. Hughes, will you come to the witness chair, please. Will you just raise your right hand and be sworn?*

Do you solemnly swear that in the matter now pending before this committee, you will tell the truth, the whole truth, and nothing but the truth, so help you God? .

Mr. Hughes. *I do.*

Hughes sat down, looking darkly at his interrogators. Batteries of microphones for seven radio stations faced him on the witness table as they did the committee at their long table covered with green felt. Since there was no dais, witness and committee faced each other at eye level.

Hughes arrives for the Senate hearings.

Senator Ferguson: *Mr. Hughes, I will try and speak as loud as I can to help you, and if you do not hear I wish that you would ask that I repeat the question. What is your full name?*

Mr Hughes: *Howard R. Hughes.*

Senator Ferguson. *And what is your address?*

Mr Hughes: *Business or residence?*

Senator Ferguson: *Give us both.*

Mr Hughes: *Well, I have several business addresses in Houston, in care of the Hughes Tool Co.; and in California at 7000 Romaine Street.*

Senator Ferguson: *Will you speak into the mike so that we can all hear you?*

Mr Hughes: *Can you hear me now?*

Senator Ferguson: *That is fine.*

Mr Hughes: *In Houston, my address is care of the Hughes Tool Co.; in California it is 7000 Romaine Street, Hollywood.*

Senator Ferguson: *You are the president of the Hughes Tool Co?*

Mr Hughes: *That is correct.*

Senator Ferguson: *And I assume the largest stockholder; is that correct?*

Mr Hughes: *That is correct, also.*

Senator Ferguson: *Is it a closed corporation, or is the stock on the market?*

Mr Hughes: *The stock is not on the market.*

Senator Ferguson: *It is a closed corporation?*

Mr Hughes: *I do not know what you mean by a closed corporation.*

Senator Ferguson: *It is held in a small group; is that correct?*

Mr Hughes: *I own all of the stock.*

Senator Ferguson: *You own all of the stock?*

Mr Hughes: *Yes.*

Senator Ferguson: *That is about as small as it can get. Now, you have several divisions of the Hughes Tool Co., Mr. Hughes?*

Mr Hughes: *I do not hear you.*

Senator Ferguson: *You have several divisions of the tool company*

Mr Hughes: *That is correct.*

Senator Ferguson: *These divisions - will you just name them for us?*

Mr Hughes: *Senator, with all of this noise and these cameras, I have trouble hearing you. Can we take the pictures first and then go on with the testimony?*

Senator Ferguson: *We will take the pictures and then the photographers can get out.*

The press moved in, frantically taking as many pictures as they could. Eventually, things calmed down, and they moved away.

Senator Ferguson: *Now, all cameras will cease and we will get down to the testimony.*

Mr Hughes: *Can we turn out the lights, please?*

Senator Ferguson: *Turn out the lights, please. It will be understood that there are to be no more camera shots. If there is anybody that has not finished, he had better take it now. All right, Mr Hughes.*

Mr Hughes: *First may I say, Senator, that the half-hour afforded me to read this previous testimony is hardly adequate since the testimony consumed several hours when it was taken and I cannot read that fast. Therefore, I request that before any of this subject matter is discussed, I be given more time to read this transcript.*

Senator Ferguson: *Well, I had not in mind asking you any specific questions about that transcript, at the present time.*

Mr Hughes: *If it does not touch on this matter, that is quite all right.*

Senator Ferguson: *I do not know as it would touch on the matter, that was quite a broad field that we have covered in your testimony. How far did you get through with the testimony?*

Mr Hughes: *I was down to page 49.*

Senator Ferguson: *That is about half way through.*

Mr Hughes: *I read that very sketchily.*

Senator Ferguson: *Do I understand that you now ask for a recess to finish reading the other part of the testimony?*

Mr Hughes: *Yes.*

Senator Ferguson: *Have you anything else that you wanted to go into, that was not covered on this, and then you could read this this evening?*

Mr Hughes: *I did not understand that question.*

Senator Ferguson: *Do you have anything else that we could go into that was not covered in this particular memorandum so that you could go into this evening and we would not lose the time?*

Mr Hughes: *Well, if you have any other material that you want to discuss with me, I would be glad to go ahead with it.*

Senator Ferguson: *Mr. Hughes, I think that this memorandum probably covered most of the case in a sketchy way, and I do not know how then we could proceed if you feel that you have not had enough time. It is not customary to furnish these memoranda of the executive session, but we thought at noon that we would do that and give you at least some time to read it.*

Mr Hughes: *I would like to be either accorded additional time to read it or have this testimony involving other matters.*

Senator Ferguson: *As I say, I do not know of any other matters that we would take up except what is covered in this memorandum.It covered both the F-11, and the cargo boat.*

Mr Hughes: *Well, then, do you want me to go ahead and finish reading it?*

Senator Ferguson: *How long do you anticipate it would require you to have read it?*

Mr Hughes: *I would say an hour. I do not want to delay your hearing here.*

The Chairman (Senator Brewster). *May I speak?*

Senator Ferguson bristled. The adversarial relationship between Ferguson and Hughes was noticable to everyone in the room. As a result, the audience became restive. They were jam-packed fifteen hundred strong into a room meant to hold about one third that number - and they noisily urged the committee to get on with it. This was the moment they had come to see. Ferguson banged his ashtray on the table with exasperation.

Senator Ferguson: *Just one moment. I must have quiet in the room. I think the Chair and the committee have been as tolerant as they can be under the circumstances. We understand that this is a public business and that people are entitled to come here and to hear the evidence, but we must have quiet and we want no remarks.*

The remarks from those in the audience were uncalled for. The witness has a right to every courtesy, and the committee has the same right, and all we ask you to do if you come here-and the officers have been very courteous in allowing more than the room would normally accommodate, and that is perfectly all right - but it must be understood that we must have it quiet,

Left to right: Thomas Black, Howard Hughes, Noah Dietrich during the 1947 Senate War Investigating Committee hearings in Washington, D.C. *(Photo: Library of Congress)*

and we want no interference with either the witness or the committee.

I have said to the witness and I now say if he does not want pictures taken they will not be taken, and they have requested him about one light, and I think that he consented to the one light, is that correct, Hughes?

Mr Hughes: *What is that?*

Senator Ferguson: *I say you have consented that we put on one light for the movie cameras?*

Mr Hughes: *That is all right.*

Senator Ferguson: *Or two, I understand. Now, Senator Brewster, you have asked to be recognized.*

The Chairman (Senator Brewster). *I think the audience is entitled to a certain amount of consideration, and I trust that no undue delicacy on the part of the committee will prevent our moving forward. There has been a very considerable delay, and I thought that I understood Mr. Hughes to say that he might take up, I thought he said, the other matter, is that correct?*

Mr Hughes: *I did not hear that. Senator, will you repeat it?*

Senator Ferguson: *Senator Brewster said that he understood you to say that you might take up the other matter.*

Mr Hughes: *Gladly, yes.*

The Chairman (Senator Brewster). *I can appreciate the committee may not want to begin an inquiry as it presents some problems, but I think that we all here full realize what the situation is and that if the committee felt it advisable to proceed with certain matters of a somewhat more personal character, it might avoid any further delay in disposing of them; and I would like to add that to the extent that those matters of a more personal character concerned with Mr. Hughes and myself are a subject of the inquiry here, I shall naturally desire to eliminate myself from exercising any of the privileges or advantages of committee membership and simply ask the opportunity to discuss them with the committee following the conclusion of that phase of Mr. Hughes' testimony.*

Senator Ferguson: *Now, if the Chair might just make an observation on the record, this hearing was begun for the purpose of looking into two contracts, one with the Defense Plant Corporation, a division of the RFC, that was known as the cargo plane contract, and the other was the F-11 or the photo-reconnaissance plane.*

It naturally was the desire of the committee to complete that. It is apparent now, as indicated on the record, that there has come into the light and into this matter a secondary matter; that is a statement or challenge as to the good faith of the committee in this investigation.

Normally a committee would not digress to allow that to be brought into the case. If that was the custom and the rule, it is certain that committees could not function, because all that would be necessary would be to raise an outside issue and it would take all of the time and attention of the committee to try an outside issue.

But, as this matter goes directly to the committee, the committee as a whole, that is the subcommittee, feels that we should go into it as it concerns the committee. I would just like, and I know every other member of the subcommittee and I say 'subcommittee' because this was not a matter of the whole committee, but a matter of the consideration of the subcommittee, that we would want to keep as near as we can to the issue, and that would

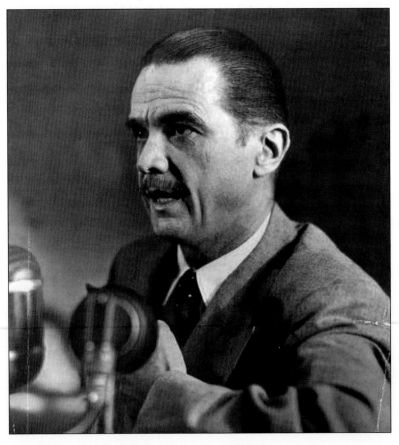

Howard Hughes
answers his inquisitors
during the Senate
hearings. *(Photo:
Library of Congress)*

be involving the committee rather than any outside extraneous matters that would have no relation to this hearing or to the matter now before us.

It is hoped by the subcommittee that this matter having come up that we would dispose of this issue now, and that we would not have it brought into the various other matters in the hearing; but we would separate it from the real reason for the committee hearing, and that was to look into the letting and the continuation and the cancellation of the two war contracts that we had in mind.

Now, I want to say to you, Mr. Hughes, do you understand what we have in mind?

Mr Hughes: *Yes, I think I do.*

Senator Ferguson: *And I am asking Senator Brewster whether the Senator has in mind what we have in mind, that while we are going to try this side issue, we are not going to make it part and parcel of the regular hearing.*

The Chairman (Senator Brewster). *That is quite agreeable with me.*

Senator Ferguson: *Not to be trying one sentence in this and another sentence in the regular hearing. So, it is your desire, is it, Mr. Hughes, at the present time, to go into the matter as to the committee at the present time, or one of its members?*

Mr Hughes: *Not my desire, but I am very willing to.*

Senator Ferguson: *There is no desire on the part of the committee to force you to go into this matter of the committee.*

Mr Hughes: *You do not have to force me. I am very happy to do it.*

Senator Ferguson: *Then you may proceed. I would ask you this. What is it that you want to tell the committee in relation to this side issue, as the committee now sees it?*

Mr Hughes: *I do not have any particular desire to tell the committee anything. I think the committee is thoroughly familiar with it. I have made certain statements to the press, and I stand by those.*

Senator Brewster has then said that he would like me to repeat those statements under oath. I have no objection to so doing ; and I think however, if I do so, Senator Brewster should take the stand and allow me to cross examine him and allow me to bring in such other witnesses as I desire, to clarify the issue.

Senator Ferguson: *Now, this has been discussed by the subcommittee and the understanding of the subcommittee is that we will apply the regular rules as far as cross examination. The subcommittee will not give to Senator Brewster the right to cross examine you nor will you have the right to cross examine Senator Brewster, but we will apply the rule that has been applied to the committee hearings ever since the chairman of the subcommittee has been on it, and I understand before that that if there are any questions you or your counsel have, if you will submit them to the committee or if Senator Brewster will submit them to the committee, then we will ask those questions, but we will do it in that way, because the same rule was applied to other witnesses and will be applied here. As to this part of the case, while Senator Brewster is a member of the whole committee, the Chair and the subcommittee as a whole feels that he should have only the same privileges that you have as a citizen under these circumstances.*

The Chairman (Senator Brewster). *Mr. Chairman, would you make it clear that I had myself stated that as my view of the propriety of these proceedings?*

Senator Ferguson: *That is correct. So that it would be clear and you would understand that you are having exactly the same rights.*

Mr Hughes: *Well, I only want the same privileges; however, it should be thoroughly understood that any questioning of me which you people carry out is in the nature of cross examination, if the questions are directed from an unfriendly standpoint. Now, I think that I should have the same privilege with respect to Senator Brewster.*

Senator Ferguson: *I think that you should wait at least until you find that they are unfriendly questions.*

Mr Hughes: *As I understand it-*

Senator Ferguson: *Do you have any unfriendly questions up to date here?*

Mr Hughes: *Not at all.* '

Senator Ferguson: *All right.*

Mr Hughes: *Then, as I understand it, you want me to tell in direct testimony approximately what happened and then the Senator is going to tell what he says happened, and then you are going to ask me certain questions, and I shall have the right to ask him; is that correct?*

Senator Ferguson: *That is right; through the committee.*

Mr Hughes: *But the questions must go through the committee?*

Senator Ferguson: *That is the rule of the committee.*

Mr Hughes: *That involves a certain delay there.*

Senator Ferguson: *A slight delay, but the committee is accustomed to that.*

Mr Hughes: *Will I have the right to call on such other witnesses as I deem desirable?*

Senator Ferguson: *We will rule on that as you request them. The committee does not want to anticipate, and when I say 'committee' I mean the subcommittee will not anticipate that question at the present time. If you name the witnesses, we will then decide that.*

Mr Hughes: *Well, Senator, I do not desire to launch into this matter and then have my hands tied.*

Senator Ferguson: *Now, Mr. Hughes, as far as the Chair can do, you will not have our hands tied. It is no desire of the committee to bring a citizen in here and tie his hands and he has not a fair hearing.*

So, as these things go on, and now you may think that we could rule, but there is no way that we can rule upon an issue unless it is presented. We are not willing as a committee to pass on hypothetical cases because they may not appear.

Mr Hughes: *It is your intention to accord me the same privileges that Senator Brewster will have?*

Senator Ferguson: *It is. That is the unanimous decision of the subcommittee.*

Mr Hughes: *You have called during the course of this hearing such witnesses as you desired, so it seems that in order to be equitable, I should have that privilege.*

Senator Ferguson *If, at the close of this hearing, you will tell us what witnesses you desire, the committee will then make a decision on that or sometime later, if you will give us the list of the witnesses that you desire.*

Mr Hughes: *Well, I am willing to proceed on that basis. What is the first step?*

Senator Ferguson: *Well, what do you have to say about this matter that we have been discussing ?*

Mr Hughes: *I think my charges have been made pretty clear in the press.*

Hughes confronts grey-haired Senator Homer Ferguson (sitting) while a smirking Owen Brewster (standing with hands in pockets) looks on. *(Photo: Library of Congress)*

Do you want me to reiterate them?

Senator Ferguson: *They are not part of the record, and it is the desire of the committee, if any statements are made, that they be made here under oath and not through the medium of the press.*

Mr Hughes: *Well, I charge specifically that during a luncheon in the Mayflower Hotel with Senator Brewster...*

Senator Ferguson: *If I interrupt, it is not to bother you at all, but I would like to get the dates as you go along and give us your personal knowledge of the facts, and then we probably can let you go right along with your conversation.*

Mr Hughes: *Senator. I am trying to be very accurate in this matter, and I am trying not to make any statement unless I am quite sure of it. Now, the luncheon was during the week commencing February 10, 1947. It was in a suite of Senator Brewster, and I charge specifically that during that luncheon the Senator in so many words told me that if I would agree to merge TWA with Pan American Airways and go along on his community airline bill, there would be no further hearing in this matter.*

Senator Ferguson: *Was that prior to the date that you testified at an executive session?*

Mr Hughes: *I believe it was afterwards.*

Senator Ferguson: *Afterwards?*

Mr Hughes: *Yes.*

Senator Ferguson: *All right. You may proceed.*

Mr Hughes: *It. must have been one or two days after it, I would say.*

Senator Ferguson: *I think that happens to be Tuesday, the 11th of February. You may proceed.*

Mr Hughes: *Now, according to this document here, testimony took place February 11; is that correct?*

Senator Ferguson: *That is correct. Look On the inside flyleaf and it will probably give you the date, also, and tell you who was present at the hearing. Would you just read the names of the committee members that were present?*

Mr Hughes: *I know who was present. Do you want me to read these?*

Senator Ferguson: *Yes. That shows the date, Tuesday, February 11, 1947, at 11 a.m., pursuant to a call, in room 457, Senate Office Building, Senator Owen Brewster, chairman of the committee, presiding; and then will you read the names of the committee members who were present?* '

Mr Hughes: *Do you want me to read these members? Senators Brewster, Ferguson, Knowland, McCarthy, Pepper, Hatch, McGrath, and O'Conner; Meader, the counsel, and Flanagan, the executive assistant.*

Senator Ferguson: *The executive assistant?*

Mr Hughes: *Yes.*

Senator Ferguson: *It was after that hearing that you went to his hotel room,*

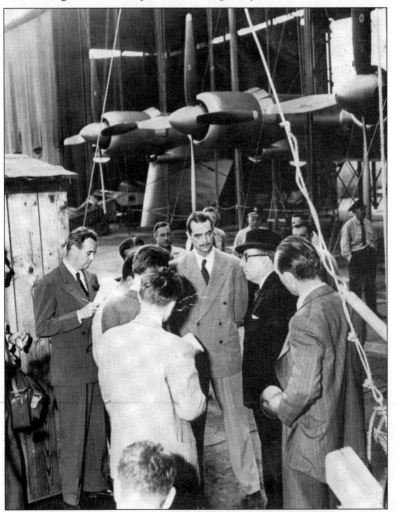

At some point at least two senators from the sub-committee visited the site on Terminal Island to be shown around the H-4 by Howard Hughes. Here Senator Claude Pepper stands alongside Howard Hughes as he speaks to members of the press.

Senator Pepper and Howard Hughes review progress on the flying boat.

you said. All right. Now, will you proceed? .

Mr Hughes: *Well, I have made my statement, Senator, do you want me to amplify it?*

Senator Ferguson: *Whatever took place, we would like to have it as full as possible, of your own knowledge.*

Mr Hughes: *I think that I have told you what took place.*

Senator Ferguson: *Now, that closes that matter, does it? Now, is there anything else that is outside this issue that you want to discuss?*

Mr Hughes: *You say is there anything further I want to say?*

Senator Ferguson: *Yes.*

Mr Hughes: *I thought Senator Brewster wanted to talk after that.*

Senator Ferguson: *Mr. Hughes, did you have a counsel at that time? A man appeared with you at the hearing, Mr. Hefron?*

Mr Hughes: *He was not a counsel. To be frank, I have been advised that Senator Brewster was very tricky, and that if I held any conversation with him alone I was taking a chance and therefore 1 should have someone with me to be sure that he could not claim I had made statements which I did not make. So I contacted a gentleman that I knew and asked if there were any in Washington who could go along with me merely for the purpose of seeing that I was not placed in an unfair position, so Mr. Hefron, whom l had never met before in my life, was suggested for that purpose.*

Senator Ferguson: *I see.*

Mr Hughes: *I took him with me first to Senator Brewster's office and then thereafter he accompanied me to the hearing; he was not my lawyer.*

Senator Ferguson: *He was not? Was he a lawyer or justice?*

Mr Hughes: *What is that?*

Senator Ferguson: *He was not a lawyer?*

Mr Hughes: *He was not my lawyer, he may be a lawyer; I believe he is a lawyer, as a matter of fact.*

Senator Ferguson: *He attended the hearing, did he not?*

Mr Hughes: *Yes, he did.*

Senator Ferguson: *And the minutes of the meeting, of the hearing, is what you have in your possession now, and that is what you want more time to read?*

Mr Hughes: *That is correct.*

Senator Ferguson: *Well, was there any extraneous matter outside of this now that you desire to put on the record?*

Mr Hughes: *Well, I don't think of any at the moment, but I would like the privilege of doing so later.*

Senator Ferguson: *All right. I would like to ask a few questions in the matter. Do you know of anything that the committee did to either carry out what you claim was asked or not to carry it out?*

Mr Hughes: *Will you repeat that?*

Senator Ferguson: *Do you know of anything that the committee did to carry out what you have claimed Senator Brewster asked, or requested, or desired, do you know of anything that the committee did to carry it out or not to carry it out?*

Mr Hughes: *Well, Senator, I have no microphone in your secret conference rooms, but I can only say that the following events transpired after this luncheon. Now during the luncheon I told Senator Brewster that I had previously discussed the matter with Juan Trippe during his dinner in his apartment at the Mayflower Hotel, that I had promised to give him my answer in 30 days.*

Thereafter, I dispatched Mr. Dietrich to Mr. Trippe in New York and asked Mr. Dietrich to find out on just what terms Mr. Trippe had in mind on this deal. Mr. Dietrich reported to me that he did not think it would be possible to make an equitable transaction with Mr. Trippe.

Senator Ferguson: *Mr. Dietrich that you are talking about is Noah Dietrich?*

Mr Hughes: *That is right, he is vice-president of my company.*

Senator Ferguson: *And Trippe is the president of Pan American?*

Mr Hughes: *Juan Trippe, yes. So I told Mr. Dietrich to drop the matter there, at least. temporarily. At just about that time, Mr. Flanagan arrived on the scene in California and started getting very tough about this investigation; up to that time everything had been very friendly. It was sort of a veiled threat.*

Mr. Flanagan hove [arrived] *on the scene and started asking all of the embarrassing questions he could think of and it was quite apparent to me that the timing was the application of screws after my refusal to go through with Mr Trippe.*

Senator Ferguson: *Mr. Hughes, I just wonder, you may not be to that part of it in the record, but was it understood or was it not understood in the record when you were there that you did not have the records but invited the committee to come out and look them over?*

Mr Hughes: *It is a strange thing, Mr. Chairman...*

Senator Ferguson: *I just wondered whether you ran into that part in the record.*

Mr Hughes: *I would like to have something to say about that. Mr.Brewster made a statement to the press that it was my own idea for Mr.Flanagan to come to the coast. It just so happened in the first 49 pages here I came across that particular portion of our discussion in the previous closed hearing, and the matter was with respect to the number of man-hours put into the design of the flying boat, I should say preliminary design, before it was submitted to the Government, and it was for such detail as that that I suggested you send someone to the coast.*

And possibly that is why Mr. Flanagan was sent, but after my failure to accede to Mr. Trippe's wishes, Mr. Flanagan of course was entirely different. He wanted to go into Johnny Meyer's expense accounts and he went down to Neil McCarthy, my ex-attorney, and said, 'Look, we know there has been a $100,000 pay-out there to Elliott Roosevelt. What about it?' And I mean his whole attitude evidently was one of making trouble.

Senator Ferguson: *Mr. Flanagan that you are speaking about is one of the investigators for the committee? Is that correct?*

Mr Hughes: *That is correct, I suppose.*

Senator Ferguson: *Did he see you personally in California?*

Mr Hughes: *Yes, he did.*

Senator Ferguson: *Did he have anyone with him at that time?*

Mr Hughes: *He might have someone with him in California, but I don't believe he had at the conference with me.*

Senator Ferguson: *When he talked with you, there was no one else.*

Mr Hughes: *I don't believe so.*

Senator Ferguson: *Is there anything else that you know of?*

Mr Hughes: *Well, to continue.*

Senator Ferguson: *By the way, about when did Mr. Flanagan get to California; have you that date?*

Mr Hughes: *What is that, Senator?*

Senator Ferguson: *The date that Mr. Flanagan came to California; do you know the date he came out there?*

Mr Hughes: *No, I don't, but I know that the timing was such in relation to Mr. Dietrich's refusal to Mr. Trippe that I felt the two were definitely connected, so then I naturally have been jealous of my reputation and there is no question but what the threat of this bad publicity was a very powerful weapon in my case.*

Consequently, and because I was pretty exhausted fighting Mr.Trippe's political manoeuvres in Washington, I talked to him on the phone and I said,

'Look, there is no chance of us getting together unless we discuss this matter in person.' And, he said that seemed to be right and would I come to New York. And I said, 'No,' that I could not come at the time, and he said well, he would come to California. So he came to California, and we had some considerable discussion and during the course of that discussion I asked him what he would do about Senator Brewster. He said that Brewster was in Europe at the time, or Asia, as the case may be. He was abroad, but he was due back in just a few days and as soon as he returned Mr.Trippe would talk to him and ask him to hold up the matter of this investigation and also would ask him to attempt a delay in the hearing of the community airline bill with the hope that we might get together on both matters.

Senator Ferguson: *How many times did you talk to Senator Brewster?*

Mr Hughes: *How many times?*

Senator Ferguson: *Yes.*

Mr Hughes: *Once in his office, once in his suite at the Mayflower, once in the airplane on the way to Columbus, and I believe once on the telephone from Los Angeles. Oh, I should mention the first conversation in Washington to Kansas City on the phone.*

Senator Harry Cain and Howard Hughes reviewing construction of the flying boat on Terminal Island.

Senator Ferguson: *Have you related or will you relate the various conversations you had with him on these various times so that we might cover this whole subject now?*

Mr Hughes: *First conversations?*

Senator Ferguson: *Yes; and follow them through at each conversation.*

Mr Hughes: *Well...*

Senator Ferguson: *Just tell us the conversations as best you can and keep them in order. The first time, and then follow through the conversations you had with Senator Brewster and later on we will ask you to follow through with those you had with Mr. Flanagan.*

Mr Hughes: *Well, Senator, if I attempted to relate all of those conversations in detail it would take an awful lot of time.*

Senator Ferguson: *No, I am just asking you the best you can, because we would like to close this incident once we put it in here.*

Mr Hughes: *Well, I should say that my first conversation with Senator Brewster when he was in Kansas City was quite brief. I understand from the telephone records it was only 3 minutes, and he said he was coming to Washington and would meet me at his apartment when he got in, but I was a little leery about that and so I told him I would see him in his office the next day. So when I met him in his office, he immediately launched into the community air-line bill. He didn't have much to say about this investigation so he talked about British supremacy, jet propulsion, and labor differential and how it was impossible for the present competitive American systems to succeed because foreign lines could supply services at the lower foreign labor rate, in direct competition with us, and I told him I did not agree.*

Furthermore, I did not agree about England's technical supremacy over us. I said that the United States in my opinion always led in commercial aviation and probably always would. I said that some foreign countries exceeded us in flash performance of various military types but that when it came to sound, safe commercial aviation this country had always led, and I felt it always would, and I felt that people might even be willing to pay a little premium to ride on the United States air line because I thought it would be safer and the public would appreciate that.

So, anyway, we had some argument on that matter and then I told him I was in Washington and while here I would like to testify before his committee if he so desired and dispose of the matter because I did not want to come back to Washington again. So...

Senator Ferguson: *Could I just inquire?*

Mr Hughes: *I am just giving you the high spots of these conversations?*

Senator Ferguson: *I understand.*

Mr Hughes: *I am not attempting to stress every point that came up.*

Senator Ferguson: *Could I inquire there whether or not the matter that you said you wanted to testify to was in relation to the cargo and the F-11, rather than the question of TWA or an other air line?*

Mr Hughes: *Well, that is correct, Senator. Some time previously I had heard that this investigation was brewing, I also heard that Pan American was behind it, and I thought rather than wait until I was subpenaed that I would offer voluntarily to testify.*

Senator Ferguson: *And it was in relation to the two contracts that you were speaking about testifying before his committee?*

Mr Hughes: *Yes.*

Senator Ferguson: *That was the War Investigating Committee ?*

Mr Hughes: *In relation to this very matter under discussion here.*

Senator Ferguson: *Well, you mean the cargo and the F-11?*

Mr Hughes: *Yes.*

Senator Ferguson: *Now, will you proceed with the conversations?*

Mr Hughes: *Well, now the conversation at the Mayflower Hotel started out with some inconsequential talk about some kind of a thing he had there in the room, it seems to me it was something hanging on the wall. Then we got around to the community airline bill again, and he gave me a lot of sales talk on that, and then we finally got into this proposition which I mentioned. Now, the conversation in the airplane was mostly...*

Senator Ferguson: *Did the airplane incident take place after the Mayflower incident?*

Mr Hughes: *Well, that I am not absolutely certain, Senator, I am trying to get some data that will place the exact date of that luncheon but the conversation in the office and my testimony before the committee and the luncheon and the airplane flight to Columbus all took place within a matter of 3 days. Which came first, in sequence, I am trying to determine. I am trying to research that and get it accurate.*

Senator Ferguson: *As I understood it previously, you had indicated that you came to the office first and then went from there immediately to the hearing room; is that correct?*

Mr Hughes: *Well, I am certain I was in the office before I was in the hearing room.*

Senator Ferguson: *And is that the first time you saw Senator Brewster in his office before you went to the hearing room!*

Mr Hughes: *I am quite certain that was the first time I saw him.*

Senator Ferguson: *Yes; all right; that was on the same day.*

Mr Hughes: *Oh, pardon me, that wasn't absolutely accurate, because he reminded me on that occasion he was one of the Senators whom I took up in the Constellation on those demonstration flights when I brought the first ship back here from the coast, but I did not remember it at the time.*

Senator Ferguson: *I see; so there was nothing happened on that occasion?*

Mr Hughes: *Oh, no, no.*

Senator Ferguson: *All right. Then, to your knowledge, that was the first meeting that you had had with him in his office and on that same day you went to the committee room and testified?*

Mr Hughes: *Well, now, that I am not sure of Senator. I believe I saw him Monday and this shows that I testified Tuesday.*

Senator Ferguson: *Then it would be the next day?*

Mr Hughes: *I presume it was.*

Senator Ferguson: *All right, you may proceed.*

Mr Hughes: *Now, in the airplane, there was more talk about the community air-line bill, Juan Trippe, and on the telephone from Los Angeles - oh yes, the conversation from Los Angeles referred to his trip to California, which he had mentioned to me at the Mayflower Hotel and in which he had offered to make in the course of this proposition that I just mentioned to you.*

Senator Ferguson: *You think - I am not quite clear on that last statement.*

Mr Hughes: *I say the conversation which I had with Senator Brewster on the phone from Los Angeles referred to his trip to California, which he had suggested making during the course of this proposition that he put to me in the Mayflower Hotel.*

Senator Ferguson: *Can you give us any more information?*

Mr Hughes: *What was that?*

Senator Ferguson: *Can you give us any more information on that as to what was said about that.*

Mr Hughes: *Where, in the Mayflower Hotel?*

Senator Ferguson: *No; about the trip to California.*

Mr Hughes: *When I talked to him on the phone?*

Senator Ferguson: *Yes.*

Mr Hughes: *Well, I think I asked him when he was coming and he said to me, he had been delayed a little while, and then I told him that I was still in the process of consideration of the matter with Trippe. I think that was about the sum and substance of it.*

Senator Ferguson: *Now is there any record that you have indicated when that telephone conversation was?*

Mr Hughes: *You mean do I have a transcript of it?*

Senator Ferguson: *No, no. Not a transcript of it, the day of it.*

Mr Hughes: *The date of it?*

Senator Ferguson: *Yes;*

Mr Hughes: *I could probably get it from my telephone records. I am not absolutely certain, because...*

Senator Ferguson: *Was it before or after Flanagan arrived out at California?*

Mr Hughes: *Well, I think I could determine that for you, Senator.*

Senator Ferguson: *Well, I wish you would.*

Mr Hughes: *I will try to do that; yes.*

Senator Ferguson: *I wish you would get that; yes. Is that what you have to say on this side issue, or the outside issue that the subcommittee has before it?*

Mr Hughes: *I think that is substantially it.*

Senator Ferguson: *That is it?*

Mr Hughes: *Yes, sir.*

Senator Ferguson then turned to Senator Brewster.

Senator Ferguson: *Senator Brewster, do you desire to take the stand?'*

Senator Brewster: *Yes ; I certainly do desire to take the stand.*

Senator Ralph Owen Brewster was not a handsome man. The forepart of his head was billiard-bald. His meaty lower lip protruded as though stuffed with chewing tobacco. Nevertheless, he carried himself as a man confidently moving in familiar surroundings and among friends. After being sworn in, he took his seat at the witness table facing the committee. Hughes sat right behind him listening intently, occasionally shaking his head or snorting in disbelief.

Brewster attempted to project an image of a reasonable man as he reviewed the history of the investigation across pages of transcript. The committee had been interested in the Hughes contracts since Harry Truman had been chairman. The committee had given the contracts time to be vindicated, but they had not been. The committee expressed concern about what had happened to the eighteen million dollars poured into the flying boat.

As to Hughes's charges against him, no, it was Hughes, not he, who had suggested flying him to his speaking engagements so that he could meet with Hughes in Washington. It was Hughes, not he, who raised the subject of TWA merging with Pan American. Except for taking a demonstration ride with Hughes in the new Lockheed Constellation several years before, Brewster had first met Hughes only two days before the alleged blackmail threat.

Senator Brewster: *It is inconceivable to me, that anyone would seriously contemplate that anyone who has been in public life as long as I have - in the state legislature, as governor, in the House and Senate - could, on such short acquaintance and in one short meeting, make so bald a proposition as he describes. It sounds more like Hollywood than Washington. No one of any competence or experience could make such a proposition. I can assure you I never did.*

Then Brewster spoke about the Fulton visit, in a voice broken by deep, theatrical emotion.

Senator Brewster: *That reveals pretty clearly that they were seeking to lay a trap for me. I promptly appointed a subcommittee headed by Senator Ferguson to handle this matter and let the chips fall where they may. I cannot and will not yield to a campaign of this character.*

It later came to light that the setting up of the subcommittee was not at Brewster's request. Rogers and Flanagan were concerned about how Brewster was handling himself. Fearing that damage to the committee's reputation might prejudice the case they were building against General Benny Meyers, they went to Ferguson and urged that he ask Senator Robert Taft, the majority leader, to remove Brewster from the committee. They thought that Brewster was going to wreck the Senate's reputation, himself, and everyone connected with him. Ferguson discussed this matter with Taft who thereupon ordered Brewster to remove himself from the committee.

After Brewster finished his statement, he was questioned briefly by Senator Ferguson in a manner that was clearly slanted to reinforce Brewster's side of the story.

Senator Ferguson: *Mr. Hughes, I wanted to ask you now-rather, first I want to ask Senator Brewster whether he has any questions for Mr. Hughes, and Mr. Hughes, do you have any questions for Senator Brewster? If so, I would like to know.'*

Mr. Hughes: *You asked me if I have any questions. Yes, somewhere between two and five hundred. Can we get started on them?*

Senator Ferguson: *I beg your pardon?*

Mr. Hughes: *I have between two and five hundred questions that I would like to ask him.*

Senator Ferguson: *Between two hundred and five hundred questions?*

Mr. Hughes: Just about.

Senator Ferguson: *When can you submit those to the committee?*

Mr. Hughes: *I will have to submit them as we go along; one would be predicated upon his answer of a previous.*

Senator Ferguson: *I will ask both sides of this controversy to submit questions to the committee chairman at nine-fifteen tomorrow morning. All questions will be in writing and submitted at that time.*

The Second Day of the Hearing

Hughes's demeanor that morning indicated that he was there to make charges, not to answer them. If Ferguson didn't exert strict control, his partial surrender of control over the hearings would be complete, for clearly both he and the dignity of the Senate were being challenged.

As the last act of the previous day's session, Ferguson had made the procedure of submitting written questions an explicit part of the record, specifying that such questions were to be submitted by 9:15. Nothing has been placed before them. Ferguson called the session to order.

Senator Ferguson: *I will now ask, notwithstanding the fact that they have not been submitted, are there any questions upon your part, Mr. Hughes? Do you have any written questions that you now desire to give to this subcommittee?'*

Mr. Hughes: *Senator Brewster, I would like to say...*

Senator Ferguson: *Just a moment.*

Mr Hughes: *Senator Ferguson, may I make a very brief statement of my feeling in this matter?*

Senator Ferguson: *I would like first to ask you this question: Do you have - and you will pardon me for speaking loud because I understand you have difficulty hearing - any questions in writing that you desire the subcommittee to ask Senator Brewster?*

Mr. Hughes: *Senator Ferguson, I have, yes, one question that I will submit to start with. Now, last evening I tried to make my position very clear that it would be impossible for me to submit in advance all of the questions that I wished to ask Senator Brewster, because each question would be predicated upon his answer of the one before.*

Senator Ferguson: *We have given the forum to you and to Senator Brewster on this particular side, and we want to be courteous and we want to have both of you feel that you have had a fair hearing on the matter, but we cannot change the rules of the committee.*

Do you have any questions to submit to the committee? This thing could go on for weeks if we are to have questions submitted in that manner, that you now ask us to do. Do you have any questions?'

Mr. Hughes: *I believe that I answered that, Senator Ferguson.*

Senator Ferguson: *Will you just pass them to the chair?*

Mr. Hughes: *Well, Senator, may I refer...*

Senator Ferguson: *Now, we are going to get this thing settled. Do you or do you not have any questions?*

Mr. Hughes: *Yes; I have some questions.'*

Senator Ferguson: *Then pass them to the chair!*

Mr. Hughes: *I will submit these questions, but I want to make it very clear that these are not all of the questions that I desire to ask Senator Brewster, and furthermore, that when I entered into this controversy yesterday I was assured that I would be allowed to cross-question through the committee, that I could ask you questions, and you in turn could ask them of Senator Brewster. Now Senator Brewster's story as related here yesterday is a pack of lies, and I can tear it apart if allowed to cross-question him, and it is unfair to place me in the position of having my integrity questioned, and not being allowed to cross-examine Senator Brewster.*

Senator Ferguson: *You understood, and the record is clear, that you could submit questions through the committee. Now, have you any questions?'*

Mr. Hughes: *Yes; but these are the preliminary questions and not all of them by any means. I cannot write the questions before I know Senator Brewster's answers. How can anyone cross-examine a witness if he is not allowed to ask questions predicated upon the answers given?'*

Ferguson held his ground.

Senator Ferguson: *The committee will rule on the matter. We are going to control the hearing.*

Mr. Hughes: *May I ask you something, Senator Ferguson?*

Senator Ferguson: *Not at the present time, until we go over these questions.*

Mr. Hughes: *May I make one very brief statement, please? Yesterday you told me that I would be accorded the same privilege as the Senator from Maine or anyone else here. Now, will you give me the questions in advance that you want to ask me while I am here, or will you give them to any third neutral party? Is it not true that you are going to propound the questions for me as I testify, and based upon what I testify as I go along? Are you willing to set aside the questions in advance which you will ask me?*

The audience erupted into loud cheers and applause. Ferguson frantically tried to regain control.

Senator Ferguson: *The officers will clear the room except for the press. We are going to run this hearing properly.*

Mr. Hughes: *I want to say that I thank you, and at least I realize the people want to see fair play here.*

Above the tumult the voice of a young college student rang out: *I speak for the audience. We will be quiet. As American citizens we would like to hear this because-...*

A number of guards pounced on the student.

Senator Ferguson, surveying the disorder with dismay, intervened.

Senator Ferguson: *He said he would be quiet. Do not remove him for saying he would be quiet.*

Photographers closed in, cameras ready, as Senator Brewster again took the stand. Flashbulbs popped and shutters clicked.

Senator Ferguson: *Just as soon as the photographers are through, the committee will come to order.'*

Ferguson began asking Hughes's written questions one by one.

Senator Ferguson: *Is it not true that you are a personal friend of Juan Trippe, president of Pan American Airways?'*

Senator Brewster: *I have known Mr. Trippe for the past four or five years perhaps. I think that I have had two dinners at which I was a guest of Mr. Trippe, once in his apartment in New York. Mrs. Brewster was present with me. And once in his country home in Connecticut where I stopped on the way home and had dinner with him.*

Senator Ferguson: *The next question submitted by Mr. Hughes: 'Is it not true that you are a close personal friend of Sam Pryor?'* (Pryor was vice president of Pan American Airways.)

Senator Brewster: *That certainly is true. I have been a close personal friend of his for a good many years.*

Senator Ferguson: *Is it not true that you are a close personal friend of Bill McEvoy, and what is his position?*

Senator Brewster: *I know Bill McEvoy, who is also a vice president of Pan American, and my associations with him have always been very pleasant, but very limited...*

Senator Ferguson: *Have you ever been a guest of Mr. Pryor, and partaken of his hospitality at his home in Florida?'*

Senator Brewster: *Twice, I think in the last two years, Mrs. Brewster and I*

*occupied for one week this small place which he has there of five rooms.'
Mr. and Mrs. Pryor were not there. We had the exclusive occupancy and I
hired a cook, whom I paid $5 a day, and I went over to the grocery store and
bought the groceries, and the Thanksgiving turkey and I left the place pretty
well stocked up, when I got through, with canned goods, as a sort of an
expression of my appreciation for what had gone on.*

Senator Ferguson: *The next question submitted by Mr. Hughes is, 'Have
you ever accepted free airplane trips from Pan American Airways in their
special private airplane?*

Senator Brewster: *I have.*

Senator Ferguson: *The next question is, 'Have you accepted such
transportation to Raleigh, N.C.?*

Senator Brewster: *I did.*

Senator Ferguson: *The next question submitted by Mr. Hughes is, 'What
was the purpose of your visit there?*

Senator Brewster's air of confidence was slowly starting to evaporate as
he realised the cumulative effects of his admissions, the smug grin on his
face was replaced by a worried look.

Senator Brewster: *I went down there to confer with Senator Bailey, who was
ill, in connection with the community airline bill which was then pending
before the committee, and it was necessary to determine what our procedure
would be.*

Senator Ferguson: *The next question submitted by Mr. Hughes: 'Have you
accepted free airplane trips in this same ship to Hobe Sound, Florida?*

Senator Brewster: *Well, I may have...*

Step by step the questions continued, slowly, but inexorably leading up
to key points from Hughes.

Senator Ferguson: *The next question submitted by Mr. Hughes is, Is it not
true, Senator, that by virtue of this powerful position as chairman of this
committee you held a whip-handle in one hand which you could easily use
to the embarrassment of Howard Hughes, while with the other hand you*

Senator Owen Brewster
at the table, with
Howard Hughes and
Noah Dietrich behind.
*(Photo: Library of
Congress)*

wanted from him his support of the community airline bill, or the removal of his opposition to it?

Senator Brewster: *No, I did not have the whip hand, nor could a chairman control the actions of the committee.*

Senator Ferguson: *The next question submitted by Mr. Hughes is, 'Now, Senator, are you going to ask the public to believe that having this unique position you did not use the whip that was in your hand to try and extract from Howard Hughes the things that you wanted so badly?'*

Senator Brewster: *That question challenges my good faith. The committee and the public will have to judge that from my record. The transcript of the March telephone conversation between Hughes and myself does not sound to me in terms of anybody cracking a whip or making threats.*

Senator Ferguson: *While occupying this unique position, did you or did you not lobby with Howard Hughes or attempt to sell him your community airline bill?' '*

Senator Brewster: *Yes, I did talk with Mr. Hughes.*

The questioning was not going well for Senator Brewster. He minimised the amount and the frequency of favours he had received and attributed them to personal friendship divorced from politics. But the picture that was emerging was of a series of close personal relationships with those who had the most to gain from Hughes's downfall: all the time while he was sponsoring a bill that would give Pan American a worldwide monopoly supported and enforced by the United States Government; and that he accepted free transportation from the organisation that would gain the most in order to promote the bill.

Hughes continued with his attack, for Ferguson had not followed up on the implications of Brewster's answers

Mr. Hughes: *I have many other questions, but I do not see any point of submitting them in this manner. I think that Senator Brewster's statements were evasive, and in many cases it was not possible to get a direct answer, and I think that submitting the questions in advance this way and having them asked and never having his reply challenged by the chairman in any way, just accepting the answers on every occasion, is unsatisfactory, so I shan't submit any further in this way.*

Senator Ferguson: *The chair has not seen fit to change the questions or to challenge them in any way. If there are no other questions, this phase of this particular hearing - the phase in relation to questions upon the part of Mr. Hughes - that phase of the hearing is closed.*

Howard Hughes' manner, refusal to accord the respect that senators ordinarily take for granted, and his attempts to run the hearings his way, had had the effect of getting under Ferguson's skin. So much so that even the suggestion of a break for lunch became abrasive.

Mr. Hughes: *May we have a slight recess for me to get my material together here? It is lunchtime anyway.*

Senator Ferguson: *Well, again, you are trying to tell the committee what it should do.*

Hughes began that afternoon's testimony.

Mr. Hughes: *The public have witnessed two men getting up here under oath and telling things which were quite contrary to one another. Now, I believe that when you buy something you generally examine the reputation of the merchant, and I think that in deciding which one of us was telling the truth yesterday our reputations should be examined at least to some degree.*

Now, I may be a little unkind in what I am going to say, but I shall also be unkind in the appraisal of my own reputation. I understand that Senator Brewster has the reputation of being clever, resourceful, a terrific public speaker, a man who can hold an audience in the palm of his hand, and that he has the reputation of being one of the greatest trick-shot artists in Washington, one of the most high-powered behind the scenes.

Now, let us examine my reputation. I am supposed to be many things which are not complimentary. I am supposed to be capricious. I have been called a playboy, and I have been called eccentric, but I do not believe that I have the reputation of being a liar. For twenty-three years nobody has questioned my word. I think my reputation in that respect meets what most Texans consider important.

Now, the statements I made on the stand here yesterday were true. I tried not to say anything which was a half-truth. I tried not to make any statements of which I was uncertain. A man who carefully and diligently tries to tell only what he is certain of, that man should be believed before a man who makes certain obvious misstatements.

Now, since the beginning of this affair, Senator Brewster has made a number of mis-statements which are obviously untrue. One of these was his statement that he has no direct or indirect connection with Pan American Airways. Now most people in the aviation industry know that statement is untrue; and in Senator Brewster's testimony here this morning, he admitted a close relationship with Sam Pryor, the vice president of Pan American.

Now, I maintain that if Senator Brewster makes one statement which is known to be untrue, there certainly should be some considerable doubt attached to the other statements he has made.

Hughes was interrupted by Senator Brewster.

Senator Brewster: M*r. Chairman - I would simply like to make a point if I may, and I do not know whether you call it a point of order.*

Senator Ferguson: *You may.*

Senator Brewster: *In what Mr. Hughes is now stating, he is reading from a prepared statement, and I think that it may be important at some point to determine whether or not these are the words of Mr. Carl Byoir, publicity manager formerly of German interests in this country and now acting as the publicity representative for Mr. Hughes. I would very much prefer to hear Mr. Hughes's testimony here rather than the voice of Mr. Byoir.*

Here Senator Claude Pepper, Democratic senator from Florida, had finally heard enough.

Senator Pepper: *Mr Chairman, I protest. Senator Brewster has no more right, since he has put himself in the position of a private individual, to interrupt Mr. Hughes's statement and to impute to him the influence of somebody else, than Mr. Hughes would have to interrupt him, which he did not do.*

Senator Ferguson let Senator Brewster's 'point' stand.

Mr. Hughes: *I will gladly state under oath that the document I have read thus far is in my own handwriting, scribbled mostly during the luncheon period.*

Later the Press made much of Senator Brewster's reference to 'German interests' as a rather under-handed way of attacking Hughes's testimony and the comment further tarnished Brewster's image.

Ferguson continued to question Hughes about his activities regarding the alleged blackmail threat, and frequently interrupted Hughes's responses.

Mr.Hughes*: Well, Senator, it is hard for me to swing from one thing to another.*

Now I am trying to make a point here, and Senator Brewster was not interrupted.'

Senator Ferguson: *Now, Mr. Hughes, I think we might as well make this thing clear right now. You are trying to discredit the committee so that it cannot properly carry out its functions . It is obvious that you are trying to take control of the Senate hearings.*

If you believe that because you have great wealth and access to certain publicity channels, and therefore you can intimidate any member of the subcommittee, I want to advise you, Mr. Hughes, that you are mistaken, and that is final . This extraneous matter, and your contempt for this committee, will not affect or cloud the real issue. This committee intends to carry out its functions. Now, is it clear?

Senator Pepper: *Just allow me, as a member of the committee, to make an observation at this point. The chairman has evidently been reading from a prepared, typewritten statement which was supposed to be elicited by an incident that occurred just before the chairman made the statement, so it must have been that the typewritten statement was prepared at some time prior to the time the incident occurred.*

Furthermore, I am authorized to say on behalf of the minority members of this committee, who are present, that he does not speak for all of the members of the committee in making the charge against the witness, and it is not necessary, probably, to resort to recrimination and countercharge properly to conduct with appropriate decorum the hearings which we are now engaging in.

Senator Ferguson: *The chair has decided, and this is definitely final, that we are not going off in a side issue which was started, Mr. Hughes, by you, many weeks ago in the press. We are not going to proceed along any such lines.*

Thomas A. Slack, Hughes's attorney, tried to get the attention of the chair.

Mr. Slack: *May I?'*

Senator Ferguson: *No, you may not*

Mr. Slack: *Thank you, Senator for your courtesy.*

Senator Ferguson: *Mr. Hughes, you may proceed.*

Mr. Hughes: *Senator, I only want to say this. Number one, I did not suggest bringing this matter into the hearing. Senator Brewster made a statement to the press.*

Senator Ferguson: *Mr. Hughes, you started it in the public, so that it would discredit this committee .*

Mr. Hughes: *I deny that.*

Senator Ferguson: *What was your motive, Mr. Hughes?*

Mr. Hughes: *I think my motive was very simple.*

Senator Ferguson: *What was it?*

Mr. Hughes: *I felt that a great injustice had been done to me, and that I should be allowed to tell my story, and tell my side of it. Before I issued those statements Senator Brewster here had made statements to the press which were absolutely false, one of which was that not even a cotter key had resulted from these contracts.*

Now, I did not ask to bring this matter into this forum. Senator Brewster asked it. I requested, before going into it, an equal status with him. Now, that has been denied, because I was telling my side of this thing here and I was interrupted, and the Senator was not interrupted.

Ferguson hotly denied that Hughes was not getting equal treatment, then Hughes brought up Brewster's allegations of a smear campaign.

Mr. Hughes: *While we are talking about smear campaigns, I wonder if it was necessary to bring in a young airline hostess, who was not in any way involved in this matter?*

The Senator stated to the press that this hostess absolutely refused to accompany us on our trip to Columbus, Ohio, because she did not dare to be alone with me in the airplane on the return flight.

Hughes then submitted affidavits from Harriet Appelwick, the air hostess, and Captain M. E. Bell, the captain on the TWA aircraft that brought the senators back from Columbus. These affidavits completely discredited Brewster's story and substantiated Hughes's assessment of him as a trick-shot artist and a liar.

Ferguson continued to try find a way of winding up what he still referred to as 'the side issue'.

Senator Ferguson: *Senator Brewster has asked for ten minutes – would you be satisfied to listen for the ten minutes: Then you would have, let us say twenty minutes?*

Mr. Hughes: *No sir. I am willing to let Senator Brewster have all day if I can cross-examine him. I feel that Senator Brewster is not telling the truth. I can cross-examine him I think I can prove it. If I do not have that right, I would like to drop the matter right at this point.*

Senator Ferguson: *This issue is closed.*

Hughes's image had been enhanced and Brewster's diminished. Hughes had called Brewster a liar, and Brewster, by his own actions and testimony,

Hughes studying some papers during recess. *(Photo: Library of Congress)*

had indicated this might be so. As Washington Post put it, the conflicting testimony had *'created the impression of a series of distortion mirrors in a Coney Island house of fun'*.

The Third Day

Friday was again hot. The crowd overflowed the caucus room and lined the corridors. Senator Brewster told newsmen that he now regretted having brought up the story about the airline hostess. He then left Washington DC – by an American Airlines airliner - to have a vacation in Maine.

Ferguson was eager to begin the real investigation.

Senator Ferguson: *You have been sworn, Mr. Hughes. Just be seated. I want to ask you some questions in relation to the flying boat contract.*

As questioning proceeded, it was clear the confrontational relationship between Senator Homer Ferguson and Howard Hughes had not been left behind. Hughes could be exasperating - he frequently could not recall matters with which it was reasonable to suppose he had been involved. He quibbled regarding the meanings of words and definitions of terms.

As the morning began, Hughes answered questions concerning his first meetings with Henry Kaiser and the attempts of Russell Birdwell (a Hollywood press agent) to sell the D-2 to the government through White House contacts. Noah Dietrich took the stand and told how he had gone to see the Air Corps chief of procurement, General Echols, regarding the D-2. Dietrich said that Echols told him that he would not do business with Howard Hughes because he did not like him. Finally, Dietrich was questioned concerning items on Birdwell's expense account led once again to Johnny Meyer.

Senator Owen Brewster seen on his way to Maine from Washington via Boston.

Senator Ferguson: *Do you know where Mr. Meyers is?*

Mr. Hughes: *Meyers ?*

Senator Ferguson: *Yes; John Meyers.*

Mr. Hughes: *Oh, Meyer. No; no, I don't.*

Senator Ferguson: *Well, he was instructed to be here, and I am just advised by counsel that he is not here, and they are unable to locate him.*

Mr. Hughes: *Senator…*

Senator Ferguson: *He works for you, does he not, Mr. Hughes?*

Mr. Hughes: *He works for my company.*

Senator Ferguson: *It may be funny to you that he is not here.*

Mr. Hughes: *I didn't laugh, Senator. Somebody laughed back there.*

Senator Ferguson: *I want to know now whether you know where he is?*

Mr. Hughes: *I don't know where he is.*

Senator Ferguson: *He is in your employ,*

and why do you not know where he is?

Mr. Hughes: *There are, Senator…*. Hughes turned in an aside to Dietrich and Slack, *'about how many thousands of people?'*

Slack: *Twenty-eight thousand.*

Mr. Hughes: *There are a lot of people in my employ. Do I know where everyone is every day?'*

As Ferguson continued bickering with Hughes, Senator Pepper tried unsuccessfully to get the floor.

Senator Pepper: *We might have an understanding that you are not the only member of this committee.*

Senator Ferguson: *At least, when I am talking I am not going to be interrupted by you!*

Well, now will you see that Mr. Meyer comes in at two o'clock?

Mr. Hughes: *What? Today?*

Senator Ferguson: *Yes.*

Mr. Hughes: *No. I don't think I will.*

Senator Ferguson: *Do you think that Mr. Meyer [conducting] business at the present time is more important than this committee hearing?*

Mr. Hughes: *Well, it is more important to my company, I can tell you that!*

Senator Ferguson: *Who in your company will know the whereabouts of Mr. Meyer at the present time?*

Mr. Hughes: *I don't know, but you have had him here for unlimited questioning, and I brought him back here twice from abroad. I don't see why I should do any more than that just to accommodate you; put him up here on the stand and make a publicity show out of it.*

Senator Ferguson: *Is that the reason he is not here?'*

Mr. Hughes: *That is not the reason. I think that is the reason you want him back!*

Finally, Senator Pepper got the floor and objected to the way Senator Ferguson was handling the matter of Johnny Meyer's absence.

Senator Pepper: *It seems to me that, if the chairman is going to be judicial in the matter, that things like that could be handled without a lot of insinuations against the witness.*

Ferguson's adversarial manner and inept questioning assisted Hughes improve his public image. Questions regarding the Kaiser-Hughes contract during the Friday morning session demonstrates the point.

Senator Ferguson*: Is it not true, Mr. Hughes, that there would be no liability for breach of contract upon the part of the Kaiser-Hughes Corporation because you did use a nonprofit corporation such as you stated in the record was a country club?'*

Mr. Hughes: *I didn't say it was a country club. Now, Senator...*

Senator Ferguson: *You referred to it as a country club charter.*

Mr. Hughes: *No, Senator, look: I feel that you have repeatedly changed my wording here and if you will read the testimony…*

Hughes was given the transcript of his own testimony where Meader had asked him to describe what he meant by a nonprofit corporation. Hughes read it out for the record: Well, it was organised in a manner similar to charitable corporations, or country clubs. Its charter specified that it was not organised for the purpose of profit

Mr. Hughes: *Mr. Chairman, I don't believe that means a country club corporation and I think your wording was intended to mislead.*

Ferguson still instisted that Hughes had described his corporation as a country club.

Mr. Hughes: *Wait a minute! I said the technical, legal corporate charter was similar to the corporate charter used by a charitable organization or a country club.*

Senator Ferguson: *right.*

Mr. Hughes: *Now, that is quite different.*

Senator Ferguson: *It is not.*

Despite Ferguson's efforts, Hughes continued to dominate the hearings and cleverly deflected testimony and allegations that would have seriously damaged a less confident witness. Senator Ferguson then brought up charges by a Defense Plant Corporation inspector that he was rarely seen at the plant.

Mr. Hughes: *He could hardly have access to the design departments of the factory. I can tell you that I designed every nut and bolt that went into this airplane. I carried out the design to a greater degree than any other man that I know in the business; in fact, I am frequently accused of going too far and not delegating enough of the work to other people.*

I worked anywhere from eighteen to twenty hours a day on this project, for between six months and one year, and this, coupled with the XF-11, the feed chute, and the other work I did during the war, resulted in me being so completely broken down physically that I was sent away for a total of seven months for a rest after the war. I do not know how anybody could have worked any harder than I did. .

If I made any mistake on this airplane it was not through neglect. It was through supervising each portion of it in too much detail; in other words, as I look back on it, if I could do the job over, I would have delegated more of the work to other people which might possibly have resulted in a faster job, but I am by nature a perfectionist, and I seem to have trouble allowing anything to go through in a half-perfect condition. So if I made any mistake it was in working too hard and in doing too much of it with my own two hands. And I can bring affidavits from everybody in my organization to back that up.

Day Four

On Saturday morning Hughes expanded on his theme.

Mr. Hughes: *Yesterday I said that I spent between eighteen and twenty hours a day for a period of between six months and a year. Now that was the concentrated, heavy design work on this plane; but from that point on I spent hours and hours every day for, oh, a period of years, on this project, and I am still spending a great deal of time on it. I put the sweat of my life into this thing. I have my reputation rolled up in it, and I have stated that if it was a failure I probably will leave this country and never come back, and I mean it.*

Senator Pepper: *Mr. Hughes, have you put your company's money and your own personal effort and prestige into the construction of this cargo plane?*

Mr. Hughes: *Well, because I so happen to believe in the future of aviation in this country, and I think this plane is a step forward.*

Senator Pepper: *Do you consider that this plane has made and will make a contribution to aeronautical science and be of value to this country?*

Mr. Hughes: *I certainly do, Senator; I hope so, at least.*

Senator Pepper: *Mr Hughes, I don't recall that while I have been in the room that anybody has given a general description of the cargo plane. Would you give it to us at the present time?*

Mr. Hughes: *Well, yes, sir; to the best of my ability. The wingspan is 320 feet. That is more than a football field, as you know; more than a city block. The length is over 200 feet. The circle on which the hull is built is twenty-four feet in diameter. It has eight engines, present rating of 3,000 horsepower each. And when gasoline turbines or fuel-oil turbines become available, then the airplane should have a great deal of power. The wing area is 11,460 square feet, which I believe is almost three times any other plane that has ever been built at this time.*

Senator Pepper then asked Hughes about the charges that there was a spirit of soaking the Defense Plant Corporation within Hughes company.

Mr. Hughes: *May I also say that there was in the neighborhood of $150,000 or more of expense which went into the airplane and which I considered entirely legitimate and which the government auditors refused to pay and which my company bore as a loss. Now, I don't believe the government got the worst of this deal, and if there are any small instances where charges were put into this project which should not have gone into it, I think there were many more instances where my company had to pay for things that went into this airplane, which the government would not approve.*

Senator Ferguson: *Well, Mr. Hughes, this matter* [the charges from the Defense Plant Corporation inspector] *that I read this morning here was a very serious charge against your company, was it not?*

Mr. Hughes: *No; I don't consider it such in any way. I think that you will find that every contract that was ever performed for the government during the war was subject to criticism of some instance where the company apparently was charging something to it which the government did not consider correct.*

In responding to the charges of inefficiency and lengthy delay in fulfilling the contract, Hughes said, 'I would like to say at this point that I have researched all of the airplanes that were designed and built during the last seven years, and I have discovered that this flying boat which is now being criticised so violently will cost the government less per pound and has taken less time per pound to complete than any other airplane, the first of a design .... The criticism that has been levelled at our supposedly inefficient operation does not really seem to be justified when you consider the result, because you can always find somebody around a factory who will bellyache and say the things are not going right, but just the same, if the result, the proof of the pudding is in the eating-if that measures up in relation to the other projects, I don't think the criticism is well founded.

Senator Pepper took his turn in questioning Hughes and with his more friendly manner, Hughes responded positively by talking at length regarding the technical considerations and the potential value of his flying boat.

Mr. Hughes: *We have discovered a great, great deal about the design and building of big*

President Truman with Hugh Fulton - he had been Executive Assistant to the U.S. Attorney, Southern District of New York, 1938-1941; Chief Counsel of the Special Committee of the Senate to Investigate the National Defense Program (Truman Committee), 1941-1944. *(Photo: Library of Congress)*

airplanes. For example, it has long been considered that the bigger the airplane is, the more efficient it is. We have discovered, and I believe it to be quite important as a discovery-we have discovered that is not the fact.

If I may be technical for just a minute. The body of the airplane becomes more efficient as it is larger. Now the reason for that is obvious. The skin area, which determines the drag, goes up as the square of the size; whereas the volume, which determines the cargo or passengers it can carry, the volume goes up as the cube.

You can see that the cube will exceed the square, and therefore the carrying capacity will be greater in relation to drag as size goes up. But on the other hand, we have discovered that wing design is quite different, and that as the wing becomes larger, it weighs more per square foot than a smaller wing. In other words, a wing of 320-foot span, built according to the same design criteria, will weigh more in relation to its size than a smaller wing.

Now, a point is reached, apparently, where the loss in wing efficiency, that is, in relation to its weight, exceeds the gain in body efficiency, and where those two lines cross, apparently, it is not desirable to build a bigger ship.

Now, this one may have actually exceeded that point, but at least we will find out. And we have already found out a good deal in that direction. Now that, of course, is of some value. I think it is of considerable value in the design of further planes.

Senator Pepper: *Are you prepared to give any general idea as to whether it is regarded as a relatively finished airplane, and relatively ready for flight?*

Hughes went on to explain that it was structurally complete and had been for a number of months, but that the control system was delaying testing of the ship. He explained the control problem in a way that impressed all who heard him with his technical competence, the magnitude of the technical difficulties and the potential that successful completion of the flying boat project would make great contributions to aeronautical science.

Mr. Hughes: *Now this airplane has crossed a barrier in size. That barrier I consider to be the one where the control system can no longer be operated by a man, even in emergency. Now, up to this time, we have had airplanes which involved a booster system, like the boost on your brake on an automobile, just to make the controls easier to operate. But if that system ever failed, the pilot was still able to operate the controls manually, in an emergency, and I believe sincerely that this is the first time an airplane has ever been designed or built which was so large that no man could possibly work the controls under any circumstances.*

So, the flight of this airplane must depend completely on the power system of those controls. That has to be so accurate that every quarter-inch movement of the control up in the cabin will be accurately duplicated in the control surface of the tail of the airplane, or on the wing.

Now, that is a tremendous problem, and we have been working on it since the start. We thought we had it licked, and we found that it was not satisfactory. The response was not quick enough, and it was not accurate enough. So we have been working overtime, and as hard as possible to lick that problem, and we think now, in our test system inside the hangar, that we have it licked, and we are now duplicating that system in the airplane, but it is a tremendous problem; and even if the airplane never flies, the research that we have done in that direction, and the knowledge we have gained, will be of considerable value to everyone in the building of bigger planes hereafter.

If any more of these airplanes are ever built they will undoubtedly be built out of metal, but I think that this airplane out of wood will serve its purpose experimentally if its other qualifications suffice.

I am not pretending to say that this airplane will fly, or that it will be successful. I only hope it will, but I think that if it is successful, the fact that it is built of wood will not preclude its experimental value, or most of its experimental value.

Now this is not an airplane that can be used to haul excursion passengers from Coney Island to Staten Island. This is not an airplane which as one article can ever be used in a commercial sense. It can only be used for testing and research and to provide knowledge which will advance the art of aviation in this country.

Hughes explained that one of his main reasons for fighting against the cancellation of the flying boat contract, was that his thinking was that for a 'comparatively small expense to the government we could finish and fly the biggest airplane in the world,' and that the value of the information derived from this would exceed the additional expenditure, when viewed in terms of what the government was paying for aeronautical research at the NACA laboratory and other places.

Under Pepper's less aggressive and biased questioning, Hughes reviewed the history of his attempts to sell the Army Air Forces a fighter aircraft based on his record-breaking racer - the plane that *Jane's* had called the *'most efficient in the world'*, but they turned it down because at that time the Army did not think a cantilever monoplane was proper for a fighter. He recounted his attempts to sell the Air Corps a high performance, two-engine interceptor with a revolutionary twin-boom tail design, and how the contract was given to Lockheed who then used the twin-boom idea.

Surrounded by microphones and documents,with the aid of an earpiece to hear what was being said, Howard Hughes answers questions about the flying boat. *(Photo: Library of Congress)*

Mr. Hughes: *When they turned down my two-engine interceptor design and gave Lockheed the contract for it when I felt Lockheed got the two-engine idea from some of my engineers, I felt I had gotten rough treatment, so I backed into my shell and decided to design and build from the ground up with my own money an entirely new airplane which would be so sensational in its performance that the Army would have to accept it.*

Now, I designed and built that under closed doors, without any really important assistance from the Army. But they didn't like that because they like to have their fingers in everything, naturally. So the Alexander incident and the General Arnold incident, plus my closed-door policy on this new airplane, left me in a position where the Army did not want to touch me with a ten-foot pole and that is why I had so much trouble getting anything at the start of the war and that is why I wound up with two contracts, half-way through the war, and such little support on priorities and other assistance that I could not possibly finish the airplanes in time to be of use and nobody else could.

Hughes' interrogation had definitely run out of steam by midday, but not before he managed to introduce into the record, over the objections of Ferguson, a news item dated 2 August 1947 quoting General Henry H 'Hap' Arnold as saying: *'Hindsight and second guessing seem to be in order. That is what some people seem to have the most of. We wanted the best photo plane we could get. That seemed to be the British Mosquito (a plywood airplane). We tried to reproduce the Mosquito in this country.*

We could not get it. We tried to do it in Canada. We couldn't. The only plane that had the prospects of equaling the Mosquito was Howard Hughes's F-11. That is why we bought it.

Hughes reviewed what he saw as the reasons why the officers at Wright Field opposed him. First, there were the incidents related to his go-it-alone,

Mrs. Claude Pepper, wife of the Senator from Florida, is shown shaking hands with Howard Hughes, as he emerged from the Senate War Investigating Committee hearing room after yesterday's session on the witness stand. Mrs. Pepper had been a constant spectator at the hearings. Standing beside Hughes is Slack, his counsel. *(Photo: Library of Congress)*

closed-doors policies.

Mr. Hughes: *Secondly, because they considered me to be stuck up. They considered that I thought I was too good for them, that I sat out in my bailiwick in Hollywood, I did not come to Wright Field to, let us say, kowtow to them there, and when they came to Hollywood, I ignored them. I did not meet them. I did not entertain them. I did not extend the type of reception they got from the other manufacturers.*

On this note, the hearings recessed at 12:45pm until 10am Monday, 11 August.

Hugh Fulton, denied a chance earlier to answer Brewster's charges of a plot in Hughes's interest, took his case to the press. Fulton had been the Chief Counsel of the Special Committee of the Senate to Investigate the National Defense Program

In a 6,000-word statement he said that: '...*Brewster was utterly unable to take issue with the fundamental charge of Hughes that while Brewster was chairman of a committee investigating Hughes and had occasion to meet Hughes only in connection with that investigation, Brewster had used that opportunity: (1) to obtain free transportation; and (2) to attempt to sell Hughes on a program to which he was known to be bitterly and publicly opposed.*

I respect the Senate Committee investigating the National Defense Program, gave four years of my life to working with the members of that

Outside the hearings, following the adjournment, Howard Hughes accepts the congratulations of the crowd, and signs a few autographs.

*committee and to developing it from an unknown committee to one of
national stature with a world-wide reputation for thoroughness and fairness.*

*I believe, therefore that as one of the chief architects of the development
of the committee, I have the right and the duty to point out to Owen Brewster
that his conduct in this matter and the false statements which he has made
in an effort to avoid the implications brought forward by Hughes have
seriously impaired public confidence in the committee under his
chairmanship.*

Day Five - Adjournment

On Monday morning Hughes entered the caucus room determined to
broaden his charges that he had to force his way into war work over the
hatred of some high Army officers. He had hundreds of pages of notes at
hand as he sat in the witness chair waiting for Homer Ferguson to call the
committee to order, whch he duly did.

But the calling to order did not pre-stage another day of enquiry – it was
just long enough for Ferguson to make a few remarks about the importance
of John Meyer's testimony and that he had not yet been found. With that as
an excuse, plus the fact that *'...all of the members of the committee have
their work to do, some are going to leave the country for a six-week period,
and other things are scheduled'*, he promptly adjourned the hearings until
the 17th day of November!

The events took Hughes completely by surprise. Smiling, he turned to
reporters to predict the hearing would never be resumed. Hughes took some
paper and swiftly wrote out a statement for the newsreels, passing it over to
his attorney, Tom Slack. *'Here, is this libelous?'*

Assured that it was not, Hughes spoke to the newsreels: *'I have just been
asked why I think the hearing was called off. I believe it was called off
because the people of this country render the final decision in any
controversy. The public is the final judge. It was very obvious from the time
I first walked into this room that the public and the press were on my side.
They believed that there was no justification for this investigation in the first
place. As soon as Senator Brewster saw he was fighting a losing battle
against public opinion he folded up and took a run-out powder. Yes, when
Senator Brewster headed for the backwoods of Maine that was the tip-off.
Washington was getting too hot for him! There was no reason for the other
Senators on this committee to continue his battle for him if he was too
cowardly to stay here and face the music. The other Senators saw no reason
to carry Senator Brewster's banner against an overwhelming avalanche of
public opinion.*

*I thought this investigation would drag my reputation through the mud.
But instead, due to the fact that the American public believes in fair play and
because they supported me, I have more friends now than I ever had in my
life. I want to thank the people of the country and the members of the press.*

Hughes's exit was barred by a cheering horde of autograph seekers
whom he forbearingly indulged for the last time. When he reached the street
to depart the Senate Office Building until November, the windows of the
building were crowded with waving spectators.

It Flies!

The Senate Hearings may have enhanced Hughes's public image, but Senator Brewster's derisive references to the flying boat as a 'flying lumberyard that would never fly' cut deep into Hughes's pride as an aircraft designer.

'I don't build cluck airplanes' he told reporters gathered around him on the day of the adjournment – now it was time to prove it. He flew back to California determined to test the plane before the hearings reconvened in November. The first thing on the agenda was to send Blandford off to Muroc Dry Lake with the XF-11 to complete the last flight tests and deliver the plane to the Air Force. This freed him to devote all his time and energy to the flying boat. His personal anger and hurt pride pushed him into super-human effort. Money was no object, he had his crews working around the clock.

'During August, September, and October, 1947 he would come in at the security gate and go over and talk with production head Jack Jerman and the plant security people there before we ever got started' recalled electrical supervisor Merle Coffee.

'We were running the engines, operating the hydraulic and electrical systems, and working on the controls at the time. I guess he was educating himself on how all the systems worked. A number of times while I was operating the electrical control panel he would stand there and ask me questions. He was trying to gain knowledge or just check me; I don't know which.'

John Glenn also recalled the time: *He came to the boat at night - he always worked at night. This man was unbelievable as far as mechanical*

Almost ready to go!

The H-4 is prepared for testing. The secondary, floating dock holds the converted PT-boat to be used as a chase vessel while the flying boat was on the water.

The view from the pilot's seat.

engineering went. He would sit there and run those engines for hours on end. Feel them out and see how they were. He knew that airplane before he flew it backside and frontside. He was unbelievable'.

Chuck Jucker, the crew chief on the flying boat saw Hughes every night. *'I'd say that ninety percent of the time he came down at night. Occasionally he brought Jean Peters down with him and sat her off in front of the flying boat in the car while he ran the engines. Jean Peters is the only one I ever saw with him, unless it was Rae Hopper, of course, or Joe Petrali.*

Hughes paid almost an obsessive attention to detail in the flight deck layout. He had sighting lines marked on the cockpit glazing marking where the horizon should be – almost an early form of head-up display. He also drove the engineers and electricians crazy in his placement of dials, switches and gauges. He'd say, 'Well, I want this switch right here'; and no matter what was behind the panel, that switch had to be right there. Placement of everything was purely for operational convenience, not for appearance or for the electrician's convenience.

Hydraulic engineer Dave Grant, who was to fly as Hughes's 'copilot' on the test flight, confirmed this: *'Howard probably spent more time positioning controls and instruments than he did anything. His cockpit arrangements were probably the best you could get'.*

Hughes seemed to understand everything that was going on. He had people quoting him on everything. He was also very meticulous and observant - he was quick to catch any variance with good aircraft practice. Hughes insisted that the men had to wear loft socks over their shoes whenever they entered the boat. He had his own rules, and if you followed them, there was no problem.

As Hughes had testified at the hearings, the development of a flight control system for such a giant machine was a major problem. But Dave Grant and his hydraulic mechanics had it nearly whipped. The system worked well in the simulator at Culver City - which simulated the static and dynamic loads on the control surfaces, duplicating the moment of inertia, the air loads, aerodynamic dampening, stiffness, and all such factors affecting flight control operation - and they were installing, hooking up, and adjusting this developed system in the flying boat itself.

By the time the hearings adjourned in mid-August, the design was completed and the simulator was only used to work out design bugs. When Hughes first returned from Washington he spent a lot of time at the Culver City simulator looking for problems and making improvements. The simulator was also used to run an endurance test of the flight control system, working up thousands of cycles in advance of the flying boat.

The flight controls were duplexed, with two hydraulic systems and Grant devised a changeover valve to switch from one to the other. This created its own problems and was only debugged just before the day of the test. In fact, engineers were doing the final adjustment on that when Hughes came aboard for the taxi test.

As 1 November approached, Hughes turned Johnny Meyer loose to create a media event that would outshine the premieres of *Hell's Angels* and *The Outlaw*. Packs of photographs of the build at Hughes Airport and the erection on Terminal Island were distributed to newspaper and magazine editors for use free of charge and Meyer invited assorted dignitaries, journalists, and photographers to Long Beach for the big event. Some, like Owen Brewster, or journalists whose editors did not think it ethical to accept Hughes's hospitality, refused. Those who did accept were wined, dined, and entertained with lavish hospitality. Hollywood stars and starlets mingled with the invited guests.

The control panels to the right of the pilot's seat. The flap lever is the handle with the cube atop it to the right of the eight throttles. The white knobbed lever between the throttles and the flap lever is the Propeller Syncroniser lever. Some sources say the pilot had only four paired throttle levers, but clearly this photograph shows there were eight individual levers.

As the flying boat was nearing completion, the Government had made preparations for the flight test programme. John Parkinson, the NACA hydrodynamicist who had worked so closely with the Hughes engineers at Langley, was a member of a committee formed to devise a test programme to justify the government's investment. Hughes had said that if the aircraft was flown, he would fly it.

NACA's chief pilot Mel Goff, who was on the committee, objected to this. He thought that Hughes lacked flight experience in flying boats, but a representative of the Reconstruction Finance Corporation who was wise in the ways of Howard Hughes laughed and said, *'Gentlemen, you can talk all day about who's going to fly the airplane in your program. But let me tell you this, Hughes is bigger than the government. If Hughes says he's going to fly the airplane, he's going to do it. There isn't anything this committee or I or anybody else can do about it.'*

In truth there was almost certainly no pilot better qualified to fly the giant flying boat than Howard Hughes. He knew all the boat's systems intimately. He had run thousands of tests in his Sikorsky S-43 in studies of

The control panel for the eight engines.

Hughes prepares to enter the H-4, accompanied by selected members of staff and media for the start of the taxi-tests, while others not so fortunate gather on the dockside to watch the flying boat be towed out.

landing and takeoff performance with various load and balance conditions. And just a week prior to the November tests he had flown one hundred twenty-six landings in one afternoon with his Sikorsky on the Colorado River behind Parker Dam as additional practice.

Saturday 1 November dawned slowly through dark clouds. Storm warnings had been hoisted from Point Conception to Newport Harbor and brisk winds stirred up choppy seas in the outer harbour at Long Beach. Hughes checked the weather and then headed for Terminal Island.

The Terminal Island workers had cleared all scaffolding and obstructions away from the flying boat. Rubber bags called 'air jacks' had been inflated

to cushion the hull as it floated in the dock. Another had been fastened under the starboard wing tip float to prevent the hull from rocking from float to float during launching.

Hughes checked these preparations when he arrived at the site. Then he walked out on the jetty that formed one side of the channel through which the hull would pass on its way to open water. Looking around he decided that tests were off for that day, but that they would continue the launching so as to be ready when the weather improved.

By mid-morning workmen had hoisted the small cofferdams from the open ends of the individual wing tip float docks and swung the massive gate sealing the main dock outward and downward so that it lay flat on the channel bottom. A large crowd of onlookers, including Paramount and Movietone newsreel cameramen, gathered at the site. People were standing all around the boat, particularly on the two jetties that formed the short channel for the hull.

All cameras were on Hughes as he arrived with his small entourage and walked past the spectators clustered on the paved surface forward of the right wing. He paused briefly, smiling dutifully before walking to the gangplank leading to the personnel entrance hatch in the nose of the flying boat below the copilot's side window. He leaned forward, put one hand on each of the wooden posts flanking the entrance to the gangplank, supported himself with his arms stiff, and swung both feet up and on to the gangplank in one agile motion.

The manoeuvrering platform on top of the flying boat.

Then he stooped to enter the hatch and climbed into the hull. Inside he climbed the ladder leading to the top of the hull at the leading edge of the wing. Here a specially designed elevator platform had been hand cranked to its raised position to serve as a vantage point for directing launching and docking operations. Jack Jerman and Joe Petrali joined him on the platform.

Who Was On Board?

Apart from Hughes himself, the other crew positions were highly coveted by everyone involved in the project. The crew chief, Chuck Jucker, made the final selections: Joe Petralli and Donald Smith, flight engineers; Warren Read, assistant chief engineer; Dave Evans, radio operator; Merle Coffee and Jack Jacobson, electricians; Thomas Dugdale and Bill Noggle, hydraulic mechanics; Phillip Thibodeau, Henry Kaiser, Al Gererink, Jim Thompson, Donald Shircy, John Glenn, Mel Glaser, and Dave Van Storm, aircraft mechanics.

In addition to the crew, several officials were aboard for the tests: George W Haldeman, the Chief, Aircraft and Components Section, CAA Region Six. He had been recently appointed by a committee of representatives of the Army, Navy, NACA, CAA, and the Reconstruction Finance Corporation to serve as technical observer of all tests of the Hughes flying boat. Mathew Whelan, Pratt and Whitney representative; Bill Newman, Hughes Company photographer; Rea Hopper, William Berry, Dave Roe, and Jim Dallas, Hughes Aircraft Company engineering representatives. A different crew listing that surfaced after the flight also records other crew members and observers on board were: Vic Leonard, hydraulics mechanic and Chris Reising, an electrical engineer.

Hughes had made provision for seats for representatives of the press, who were selected by a lottery system. The winners: Jim Padgett,

International News Service; Joe Johnson, newsreel pool photographer, and his assistant, Dexter Alley; Ralph Dighton, Associated Press; John Vonderheide, United Press International; and James McNamara, radio pool announcer (accompanied by his recording engineer). The reporters left behind were reduced to pushing and shoving for position on the press boats Hughes provided.

Hughes talks to Merle Coffee, the radio operator as the engine engineer looks on.

Once on board Hughes carefully checked everything visually, and with a walkie-talkie made contact with Harold Tegart, the launch engineer, who would direct the winch operation. Four Caterpillar tractors tended the lines that would keep the tail from swinging during launching. Everything looked good.

He gave the order to start to ease the huge flying boat out of the dock. Electric winches on shore whined, pulling in the two lines that ran from the nose of the flying boat through pulleys anchored to the harbour bottom fifty feet apart some three hundred feet offshore and marked by two black and yellow striped rubber buoys. But as the aircraft started to move for the first time into the channel between the finger jetties, a freshening wind pushed against the towering tail. The fore and aft restraining lines, tended by the D-6 and D-8 Caterpillar tractors, tightened.

Everything worked as planned except for one moment when the wind threatened to swing the tail against one of the jetties. Hughes thought during that moment that the tail was getting away from them, but the restraining line tightened and held. The aircraft cleared the end of the jetties and was winched to a stop at the buoys, where the crew of a small Hughes craft made it fast to the mooring bridle for the following day's test.

Early on Sunday morning, 2 November, Glenn Odekirk picked Hughes up in Bel Air. According to Odekirk, this was the conversation:

Hughes: *How's the weather at Long Beach?*

Odekirk: *It's not good, but it's better.*

As they drove to Terminal Island, Hughes asked *'Odie, you don't mind if you're not aboard while I'm doing my taxi tests, do you?'*

Odekirk: *Oh, come on, don't give me that bullshit. I know if it feels right you're going to hop it.'*

Hughes just smiled.

Odekirk knew that Hughes didn't want another pilot aboard. That way no one could say he did not fly it, the other guy did!

At Terminal Island, Rae Hopper talked with Hughes before they boarded the flying boat. *'We found a stress error in the aileron operating mechanism, nothing too large and easy to fix, but it would be better not to fly until we do. If you want to go ahead and fly it, don't go over one hundred forty miles per hour.'*

Hughes agreed that he would not.

Hughes visited the press tent before boarding, checking that his instructions to take good care of the press had been followed. Rows of tables covered with heavy white paper filled half the tent. One such row supported batteries of telephones, some with hand-held receivers and megaphone mouth pieces, so that reporters could flash their stories to impatient editors. On other tables sat rows of Underwood typewriters, each with a folding chair in position ready for use. There was also food and drink, catered by Hughes's friend, Hollywood restaurant owner Dave Chasen.

Around midday, Hughes invited the press to board the flying boat for what had been described as the maiden taxi test - a three-mile run up the bay and back. Hughes's small craft ferried them the short distance to where the flying boat swung tethered to the buoys.

As his guests came aboard through the forward hatch, Howard cautioned

The flying boat is slowly eased out of it's dock on Terminal Island.

them not to expect too much. 'We'll reach speeds of only about forty miles per hour. The water is too choppy for anything else.

Then he clambered up the ladder to his topside vantage point to supervise the positioning of two sea mules for towing. He then returned to the flight deck, where a number of passenger seats had been installed.

He eased himself into the pilot's seat alongside his 'co-pilot' Dave Grant. That Grant, a non-flier, should be copilot during the test programme was an idea unique to Hughes. Hughes did not want a regular copilot; with himself in full control it could never be claimed that he himself had not actually flown the big machine. Moreover, Grant was the designer of the flight control hydraulic system, regarded the most critical part of the aircraft at this time; in Hughes's mind it was fitting that Grant shared a seat at the controls.

Chuck Jucker, the crew chief, positioned himself between the two pilots, belted himself to a stool secured to the deck, and adjusted his interphone headset. Jucker was Hughes's link with the rest of the crew throughout the ship. Chief engineer Rae Hopper and program manager Bill Berry were behind Hughes on the flight deck.

The others all took their places. Flight engineers Joe Petrali and Don Smith were checking their controls and instruments just behind Grant, while Merle Coffee stood alongside the engineers and monitored the electrical panel. Across from them, just aft of the radioman's position, electrical engineer Jim Dallas monitored the control system strain gauges he had designed and installed.

Further aft on the flight deck, power plant mechanic John Glenn stood by the auxiliary electrical unit. Systems mechanic David Van Storm also was on the flight deck, but moved down to the cargo compartment when

As the flying boat was slowly eased out of the dock Howard Hughes took up a position on a manoeuvring platform atop the boat, just forward of the wing. From here he could see the 'sea mules' towing it out, and the caterpillar tractors steadying the tail.

the aircraft took off.

Electrical mechanic Jack Jacobsen manned his station forward on the cargo deck at the starboard hatch, while Ben Jiminez stood by on the port side. Hydraulic mechanic Bill Noggle moved about checking the operation of the hydraulic system and the flight controls. Power plant mechanic Al Geverink was in the right wing behind the engines and Phil Thibodeau in the left wing behind the number two engine.

Hughes had walkie-talkies on his lap for communicating with press boats, sea mules, and other craft in his fleet. After checking that everyone was on board, he picked up one and instructed the sea mules to take the flying boat in tow. One took the nose line while the other tended a line to the tail to keep it from swinging with the wind. Mooring lines were cast off. At the last moment, Ben Jiminez at the port cargo door unhooked the ship-to-shore intercom line, and they were free to move.

On the way to the outer harbour test area they kept a sharp lookout for boats

and other floating hazards. In the copilot's seat Dave Grant had no interphone headset. He had asked for one when he came aboard, but in the confusion his request had been forgotten.

Halfway out of the dock - seaworthy and ready for flight!

As the two sea mules manoeuvred the big aircraft toward the outer harbour Hughes could see Navy ships at the Long Beach Naval Base pier and a large number of small craft in the harbor. Spectators lined the breakwater at the opening to the outer harbour, waving and cheering them on. In the outer harbour an armada of water taxis and pleasure craft had gathered for the event. It was Sunday and the crowds were in a festive mood. Many of Howard's friends were watching from the yachts of the Hollywood great and near-great, and many a glass was raised in his direction.

As the big machine approached the test area, everyone was concerned that the test area was clean of floaters and other debris. That was one thing they did not need: to hit a floating timber or railroad tie at a hundred knots. A Coast Guard cutter assisted Hughes' craft in clearing the channel and in keeping the small craft out of the way.

When they arrived at the test area, Hughes and his engineers started the eight engines one by one. The sea mules finally cast off and retreated to one side.

When Hughes saw that the way ahead was clear, he advanced the throttles. As the engines increased in power to that called for by Hughes, Petrali and Smith operated their vernier throttles to even out the readings across the board. But just as they had done so Hughes hauled his throttles back again. They slowed and settled back in the water.

Hughes immediately noticed that not all engines had returned to idle. *'What's wrong with the throttles?'* he called. Petrali and Smith hastily reset their vernier throttles and the engines idled evenly again.

Thus the first run down the bay was actually a series of about three short taxi runs in which Hughes accelerated to about forty knots and then decelerated as he felt the aircraft out while moving in one continuous line. During this first slow-speed pass down the bay, Hughes had everyone in the cockpit peering through the windshield looking for logs and other floating hazards. At the far end of the bay he throttled back again and swung the aircraft around for the return trip. The flying boat handled very well on the water, responding quickly to rudder and engines.

In position for the second run, Hughes's right hand advanced the throttles again while his left hand steadied the wheel on the control column. The eight R-4360 Wasp Majors roared and the flying boat surged ahead and climbed rapidly onto its step. As the airspeed climbed past fifty, sixty, seventy to seventy-five knots, the hull spanked only the tops of the waves. Throughout the run, despite the chop which was just short of developing whitecaps, the hull spray fell away well below the propeller line.

At the end of the bay where they had started, Hughes closed the throttles with satisfaction. The run had felt good.

As the flying boat weather-cocked into the wind, engines idling, Hughes answered reporters' questions. Yes, he was satisfied with the ship's performance. Yes, according to plan, he would try one more taxi run. Yes, he felt that the seven million dollars of his own money that he had ploughed into the airplane was well spent.

John Vonderheide, the United Press correspondent asked if he was going to fly the boat this day. *'Of course not, as I have explained, I estimate it will be March or April before we're really ready to fly this airplane.*

'In that case,' asked the UP reporter, *'is there a possibility of going ashore to file my story? 'Certainly'* Hughes said. He picked up his walkie-talkie and called for a boat to come alongside.

The UP reporter's request triggered off a mass exodus. No one wanted to be scooped to the story by others. Hughes accompanied by mechanic Van Storm went down to the cargo deck to see them off.

Van Storm lay on his belly at the left forward hatch to hold the boat clear of the aircraft while two reporters walked over his back to make the transfer. *'Be careful,'* Hughes cautioned. They thought he was being solicitous of their welfare. Actually, he was thinking of the flying boat.'

Two or three magazine writers, who had no deadlines, about five other reporters, and a couple of photographers decided to stay aboard. James C. McNamara, news editor of independent Los Angeles radio station KLAC who was the 'broadcast pooler' for all the radio newsmen, had no alternative but to stay too. He had waited until the third run to wrap up his story because he thought he would gain a more intelligent approach after the on-the-spot education provided by the first two runs. By the time he could finish the third and last run and get his recording to radio-central via Long Beach station KFOX, the icing would be off the story. With a heavy heart he watched his colleagues leave.

While all this was going on Dave Grant sent Bill Noggle back to check the control system actuators, control valves, and plumbing in the tail. Hughes was

With eight engines at little above idle speed, the huge flying boat barely moves through the water.

unaware of this; neither did he make any check to see that crew and passengers were in their places and ready to go before the next and final run.

After the boat headed back to Terminal Island with the reporters, Hughes restarted all engines and taxied farther downwind to get in position. The final run would be westward toward San Pedro and roughly parallel the shoreline from Pier A in Long Beach to the western boundary of the Terminal Island Navy Base. McNamara stood just behind crew chief Chuck Jucker - a good vantage point from which to relay to the American public his view of what was about to happen.

In 1947, the reporting of special events away from the broadcasting studio was not easy. McNamara's equipment was portable, but ponderous; a storage battery for power, a heavy turntable recorder, a supply of 15-inch acetate discs, assorted cables, and a microphone-a back breaker. While McNamara told the story through his KLAC microphone, his engineer, Harold Huntzman, in a front row seat, monitored the recorder at his feet.
McNamara began his famous broadcast, which later won the 1947 Sigma Delta Chi award for the best on-the-spot reporting. *'This is James McNamara speaking to you from aboard the Howard Hughes two hundred-ton flying boat, the world's largest aircraft. At this moment as we speak to you from the spacious flight deck, this mighty monster of the skies is slowly cruising along a northwest course in the outer Los Angeles Harbor.*

And in just a moment he will make a second taxi-run, this one I believe will be for the edification of some photographers who are out here, the national publications and newsreel cameras and they will be shooting this next run. And no doubt Mr Hughes will again challenge the ninety mile (an hour) mark he set here just five or ten minutes ago. He may exceed that this time so the boys really get a great picture.

Howard, do you have time at this moment to explain for our radio listeners exactly what you're doing-the operation?'

Yeah. We're taxiing downwind very slowly to get into position for a run between the entrance to Long Beach Harbor and Myrtle Beach and San Pedro. It's about a three-mile stretch there and we're gonna make a high-

speed run in that direction. The wind is changeable-it's been changing all day-but it's not too serious.

The water is fairly rough for the outer harbor here. You see, there was quite a heavy wind last night, up to forty-five miles an hour, I believe, which created quite a sea here and that shortens the length of runway we have available because down toward the southeast end here there's effectively an open sea coming around the end of the breakwater and it's very, very rough. But I believe we have an area here that's sufficient and pretty soon we'll be ready to make a run.

McNamara was struggling to hold a conversation with Hughes. *'Howard, I didn't hear you. There's so much noise here. But did you tell our listeners the speed you achieved on that last run?'*

That was right around ninety miles an hour. We were well up on the step and the airplane could have lifted off easily if I'd just pulled back on the control, I'm sure. But we have so many mechanical devices in this ship right now that I wanta check a few of 'em out a little bit before we try anything like that and as I say, I previously estimated it'd be March or April before we're really ready to fly this airplane. You have no idea how much machinery, how many electrical, hydraulical, and mechanical devices we have on here, all which should function right before we take the ship up in the air.'

Hughes turned his attention back to the cockpit. Looking at Grant he said, *'Lower fifteen degrees of flaps.'*

This was the takeoff setting for the semi-fowler-type wing flaps. So as Grant's left hand grasped the cube-shaped knob of the flap control lever protruding from a diagonal slot in the top of the pedestal between the two men, he thought that Howard might just lift off on the next run! Grant remembered the high-speed taxi tests of the XF-11 when Hughes had 'taxied' it into the air several times well before the official test programme had started.

Grant moved the flap lever to takeoff flap position, but nothing happened. It seems that a malfunctioning hydrofuse should have been wired in the open position because this leak protector had previously shut off hydraulic actuating fluid when no leaks existed. Tom Dugdale, a

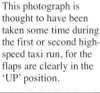

This photograph is thought to have been taken some time during the first or second high-speed taxi run, for the flaps are clearly in the 'UP' position.

hydraulicsman, quickly pulled the manual override, and the flaps moved down to the take-off position.

Hughes lined the flying boat lined up in position. The fifteen to twenty knot wind had shifted around so that the channel that had been cleared of debris and pleasure craft by the Coast Guard cutter and Hughes's boats was no longer directly into the wind.

He picked up his mike and checked if the engines were ready for another run. Receiving an affirmative, Hughes hung up his mike and spoke over his shoulder to McNamara.

'Hang on.' He made a quick visual check of cockpit instruments and control settings, looked ahead and saw that their intended path was clear. With the heel of his right hand he eased all four paired throttles forward in one smooth motion.

'This is James McNamara speaking to you from the cabin of the giant Hughes aircraft once again. Howard Hughes has just alerted us he's asked everyone to hold on as he gets all eight engines moving – tremendous horsepower picking up!'

The flying boat surged forward, engine noise mingling with the out-of-sync beat note from props not turning at exactly the same speed. The noise reverberated through the hull. The thump of the bow against the waves sounded like a speed boat's pounding in choppy water, but magnified and booming hollowly.

Then a power boat raced across their path from right to left. Hughes

swerved but the incident was more irritating than hazardous. The machine responded beautifully.

'Howard is sitting directly in front of us – I'm holding on to his seat. Here we go!'

The nose of the aircraft came up during initial acceleration until, smoothly and without effort, they were on the step. McNamara called out the speeds for the benefit of the radio audience. *'The airspeed indicator. It's twenty-five – it's thirty. It's thirty five as he pushes the throttle. It's forty. More throttle. It's forty five. More throttle. It's fifty. It's fifty over a choppy sea. It's fifty-five; it's fifty-five. More throttle. It's sixty. It bounces to sixty-five. It's seventy. It's seventy-five... And something momentarily cuts out... I think we are airborne!*

The flying boat left the water shortly after reaching seventy knots. Hughes pulled the throttles back and then immediately pushed them forward again. A change in sound that was captured on the McNamara recording.

The right wing started to drop in the cross wind. Hughes immediately wracked his wheel to the left. He was not yet sure how fast the aileron would respond so he purposely over-controlled. It was more than enough. The wing came right up and he returned the wheel to neutral.

McNamara: *'We are airborne ladies and gentlemen. I don't believe that Howard Hughes meant this to be. I don't know. We were airborne for just a moment and we were really up in the air. You could tell we were really up in the air! I don't know whether Howard...'*

The start or end of the third taxi-run - the flaps are clearly angled down into the 15 degree take off position. As with all the runs, the flying boat was shadowed by a helicopter.

After dropping off a number of members of the press, 24,000 horsepower comes back to life.

Howard has a final talk with the crew. Just visible on the flight-deck glass are the painted marks showing where the horizon was expected to be at certain times of the taxi trials, and at later stages of testing during take-off and landing.

McNamara put his microphone in front of Howard Hughes. *'Howard did you expect that? 'Exactly..' replied Hughes with a chuckle in his voice. 'I like to make surprises' 'You were surprised?' 'No, I said I thought I would make a surprise!' 'You certainly surprised the people here! The airspeed was eighty miles an hour when it took off and we were about thirty feet above the floor of the sea – I mean the surface of the sea rather, and that was certainly the greatest surprise.*

With his right hand he held the power steady for a moment. Then he pulled the throttles back, easing the flying boat down to a power-on landing. He must have been unsure of elevator response, because all the way down his left hand pumped the control column back and forth as much as plus-or-minus twenty degrees, according to copilot Grant.

The landing was so smooth, in fact, that the stylus of McNamara's recorder never even bounced.

McNamara: *'Ladies and Gentlemen, ladies and gentlemen the Hughes mammoth aircraft has flown this afternoon having negotiated about three hundred feet – I don't know what the expression was on my face, but I looked over here at the assistant engineer and I have never seen a more surprised character in all my life!*

McNamara summed up his impression for the radio audience. *'At one time, Howard said that if this ship did not fly that he would leave the country. Well, it certainly looks at this moment that Howard Hughes will be around the United States for quite some time to come.'*

James C McMamara, the well-known news editor for Los Angles radio station KLAC as he interviews Howard Hughes immediately after 'the flight'.

Then, McNamara spoke to Hughes again. *'Howard, may I again offer my congratulations, sir?'*

'Well, thank you very much.'

As McNamara signed off with his radio audience, Hughes said, *'Well, you got the first ride.'*

The H-4, escorted by attendant motor-boats off Ocean Boulevard, Long Beach, with the National Auditorium and Rainbow Pier.

Others views of 'the flight'

Before the start of the third run, Grant sent Bill Noggle to the tail. At first it was smooth and quiet there and he could hear the hiss of the hydraulic servo valves, which were designed to bypass about a gallon of fluid per minute when the controls were in neutral. Noggle had no warning of the takeoff and was halfway up the ladder to the elevator control units when it started. A tremendous shaking pinned him to the ladder. Noggle should not have been there, but he was due to Hughes' lack of operating procedure.

Harry Kaiser was on the flight deck at lift-off, but he went down on the cargo deck after touchdown and saw the tail twisting around. Noggle came

running *out 'Look at that son-of-a-bitch Harry, it's about ready to leave us.'* Ben Jiminez saw Noggle come out of the tail. *'He came down out of that empennage like he was on fire.'*

Noggle is on record saying that *'...when the plane was airborne, everything smoothed out. So I scrambled up the rest of the ladder and was able to look out and see that we were in the air.'*

Thousands of specators lined the breakwater in Long Beach harbour to watch the test taxies of the flying boat.

Jack Jacobsen, at his station in the fore part of the cargo deck, looked back and saw the long catwalk weaving and shaking: *'You could hear the wood creaking, but once we were off the water, everything just smoothed out. I could tell the minute it left the water. When it settled in the water it*

220

started creaking again'.

Dave Van Storm, who was in the port bow entry area at this time, heard no wood creaking, only engine and water noise. He thought that the movement of the tail seemed a designed movement not out of line with a designed wing deflection of seventeen feet at the tip.

Phil Thibodeau was in the port wing behind the number two engine. *'There are drain holes in the bottom of the nacelle about one-half inch big. Straight down was the only view I had. But I could sure tell when we left the water. It was very rough that day. On the water it swayed back and forth - a sidewise motion. After he lifted it, he throttled back and there was a glide. It was a real sensation.'*

Tom Dugdale, inspecting hydraulic units on the rear spar, he heard the engines rev up he figured that he was right over the centre of gravity, so he just stayed there.

Power plant mechanic Al Geverink was in the right wing behind the engines. *'They had air throttles at that time. We helped to start them - and made sure they operated properly. We took the lens out of the landing light so we could see when we were airborne.'* Power plant mechanic Don Shirley shared the view with Geverink.

Geverink was not aware of anything that suggested that Hughes had planned to fly the H-4 that day and was surprised to be airborne. *'I think he had intentions of just picking it up a bit. But once he got it up, he didn't know what to do with it. It got quieter when we got airborne. You could hear the sound of the water spanking the hull before that.'*

Ben Jiminez was also surprised: *'I wasn't even thinking of going into the air. It's just my opinion that he got on the step there and it just took off. After it took off, Hughes just rode it out.'*

Given the amount of press representatives and photographers on board for the taxi tests, it is not surprising there is a large quantity of pictures available.

However, here is strong evidence to suggest that some were posed and taken before or after the event.

In some, like the one below Hughes is wearing his coat and hat...

... where in others he is hatless and in shirtsleeves, for example, when he was being interviewed at the end of the third run by James C McMamara.

Suspiciously also, the flight deck glazing appears to have been covered over in some shots.

Others show the use of sophisticated concealed lighting to illuminate the interior - which suggests a dedicated photo-shoot either before of after the taxi tests.

Flight engineer Don Smith thought the same: *'I'm sure it ballooned out of the water before he was ready for it. The thing that convinced me is that as soon as it came off the water he yanked the throttles clear back.'*

Other crew members claimed they were aware that Hughes was going to take off. Jack Jacobsen said that before each run he cracked the forward door open enough to peek out at the wing flap position. Before the last run he saw that the flaps were set to the takeoff position.

Kaiser was aware of it too. *'When he dumped the passengers off and taxied around a little bit, we knew we were going to fly - He didn't tell us.'*

Glenn Odekirk, Carl Babberger, a number of photographers, a rescue team in fireproof suits, and a select group of passengers saw the takeoff from the flagship of the Hughes fleet, a former three-engine Navy PT boat capable of about sixty-five knots. As the flying boat roared across the outer harbour, the PT boat paralleled its course on the seaward side. The camera crew filmed the liftoff and the landing from the PT boat's stern.

After the flying boat had touched down again, the regional chief of the Civil Aeronautics Authority who was on the PT boat put his arm on Odekirk's shoulder and said, *'I knew the son of a bitch was going to do that.'*

Hughes water-taxied the flying boat back toward Terminal Island. This time he didn't bother with 'sea mules', he just taxied the flying boat in under its own power right up to the buoys as though he were maneuvering a yacht. Accounts say he talked enthusiastically about how good he thought the flying boat was. He was filled with euphoria; saying to all who would listen that the aircraft performed 'beautifully,' and several times he mentioned how the flaps had 'ballooned it.'

After the flying boat was fast to the buoys, Hughes left his seat and put on his jacket and new brown snap-brim hat, which he had apparently purchased for the occasion. On shore Hughes could not keep from grinning, and he swaggered a bit as he walked along the jetty. He chewed a sandwich as spectators clustered around him offering congratulations.

'I think the airplane is going to be fairly successful, I sort of hoped it would fly but didn't want to predict it would and make people disappointed.' Hughes's description of the takeoff: '...when I got going with the flaps in takeoff position there was so much buoyancy and it felt so good I just pulled it off. The maximum speed during the flight was a little over a hundred. The takeoff speed was ninety-five miles an hour [about eighty-two knots]. *The takeoff distance was two-fifths of a mile.*

'The landing was really gratifying. As soon as the ship was in the air I cut back the power because I didn't want to hit those small boats in San Pedro harbor. The controls operated well. We will follow the original testing schedule.'

That schedule called for a first flight in March 1948.

Hughes congratulated members of his staff who helped him design the plane and work out its technical problems. He then escorted a party of visitors aboard the flying boat via one of his small boats. Hughes explained that the flying boat was three times the size of the Martin Mars – another US Navy flying boat - and seven times the size of the Douglas DC-6, the largest commercial airliner then operating. The Hughes boat could carry an estimated 700 passengers, he told the visitors. He explained that from the cargo deck to the flight deck there is a ladder with seventeen standard steps and that the cockpit was more than thirty feet above the water line.

IT FLIES!

Did Hughes Mean To Fly It?

The question of did Hughes really intend to fly on the day set for the taxi test has been asked many times, but since hs death in 1976, is almost impossible to answer - No one really knows what was going on in his head, and given Hughes' somewhat undisciplined approach to test flying, it's almost impossible to estimate.

All the advance publicity carefully stipulated there would be only taxi tests on 1 November, but many felt this was just opening he needed in case he didn't fly. In fact, it has been said that he denied he was going to fly it, but it could be seen as him just covering his options.

From accounts of Hughes' previous test flights - the Sikorsky test on Lake Mead, the XF-11, and the flying boat he was highly informal about it all. It was his habit not to use check lists, file flight plans, brief his crew and passengers, make communications checks, or coordinate with or effectively use any crew members other than his flight engineers. The departure from Terminal Island for the day's tests, for example, was strictly informal. He did not introduce the crew to the press and did not even brief them.

Except with his engineers, Hughes made no checks with crew stations at all before taxiing or taking off, not even through his crew chief. It seems that he spent more time placing the press boat than he did anything else.

As for safety precautions, there were no life vests or life rafts readily available, nor did Hughes give safety briefings to passengers or crew. Passengers wandered around the flight deck during all phases of the test including the takeoff.

However, Merle Coffee recalled that on the night before the flight, Hughes said to those gathered on the flight deck, *'Everything looks good*

and we're going to take it out and fly it.'

But crew chief Chuck Jucker, who rode between the pilots' seats during the taxi tests and lift-off, answers the question by saying, *'Based on my instructions, I should say no. I was told to get the ship ready for taxi tests only.'*

According to Odekirk, Hughes thought the flying boat felt 'real good' during the test runs and flight. 'He was very happy about it.'

At one point after the flight, Hopper asked Hughes, *'Did you mean to take off on that run?'*

'What do you think?' retorted Hughes.

Dave Grant thought the takeoff was both planned and inadvertent. *'At the very beginning I think he really did plan to just taxi the aircraft. But he wanted to fly it and in his mind that possibility was there. However, he probably didn't really commit himself to it until after those first two runs.'*

As to the inadvertent part, Grant said that while they were designing the controls, Howard wanted a lot more response than he needed. 'It seems that all the time *'...he kept telling me how it was necessary to rock flying boats off the water and things of that type. But this particular hull was different from the others. It had absolutely no porpoising tendencies. It had excellent water handling characteristics.'*

'On previous runs we'd been at higher airspeeds without taking off. I think what actually happened was he didn't expect the added lift with only fifteen degrees of flaps. They're tremendous flaps-semi-fowler, in effect, because their hangers are way below the wing.

'With the flaps at fifteen degrees, the airplane took off as soon as it got on the step and I think he expected to have to pull it off. That's what I mean by both inadvertent and on purpose. The only thing he was surprised about was the airplane took off without him having to do it.'

The flying boat performance curves show with the flaps set at fifteen degrees, the stalling speed plotted against aircraft weight. At their test weight of about 280,000 pounds, the curves show that stall speed should be between

Howard Hughes and some of the crew pose for this 'photo opportunity after the taxi-test runs

FLIGHT CREW

HOWARD HUGHES
Pilot

D. GRANT
Copilot

C. JUCKER
Crew Chief

D. SMITH
Flight Engineer

J. PETRALLI
Flight Engineer

M. COFFEE
Radio Operator

M. THIBODEAU
Systems Mechanic

M. GLASER
Engine Mechanic

W. NOGGLE
Hydraulic Mechanic

T. DUGDALE
Hydraulic Mechanic

2 November 1947

D. SHIREY
Engine Mechanic

J. GLENN
Engine Mechanic

V. LEONARD
Hydraulic Mechanic

A. GEVERINK
Engine Mechanic

V. STORM
Systems Mechanic

J. JACOBSON
Electrical Mechanic

H. KAISER
Engine Mechanic

B. JIMONEE

Management and Engineering

R. ROE

A. BERRY

V. HOFFER

H. REED

D. EVANS

C. DALLAS

There is some mystery as to who exactly was on board the H-4 on 2 November. Different lists over the years, but possibly the most accurate is the 'list' on the commemorative photo.

It shows eighteen flight crew and six management and engineering staff.

seventy and seventy-five knots – and that is the estimated speed when the flying boat got into the air, so the HK-4 performed just as designed.

Carl Babberger is on record of saying that '...*all the factors were present for takeoff-a high head wind, the fifteen-degree flap setting, and a light load. It probably got airborne before he* [Hughes] *expected it to, but on the other hand it wouldn't surprise me that being under fire from Senator Brewster, he was prepared to gamble. If it took off, fine. If it didn't, fine.'*

Aftermath

After the successful flight of the flying boat everyone understood *Time* magazine's terse headline: 'It Flies!' But after the first flush of publicity it seemed just as Grover Loening had predicted: 'I think it will fly... and that after a great many pictures have been taken of the crew and pilot, with Mr. Hughes looking very tired and very heroic - it would probably be run up on the beach and stay there, like any other movie set.'

The November Senate hearing was something of an anticlimax. Senator Brewster had gone after big game – Howard Hughes and the supposed Roosevelt connection. Instead, the committee bagged Brig. Gen. Bennett 'Benny' Meyers, whose financial misdeeds were uncovered during the Hughes investigation.

Meyers had established the Budget Office in 1935 at Wright Field, Ohio and was budget officer and chief of that division until September 1940 when he was transferred to the Office of the Chief of Air Corps as assistant executive. He became executive officer of the Materiel Command in that

Office in November 1940, and in March 1942 was named deputy to the Assistant Chief of Air Staff of the Army Air Forces. He assumed command of the Materiel Command, with headquarters at Wright Field, Ohio, in June 1944, and the following month was named Deputy Director, Army Air Forces Materiel and Services at Patterson Field, Ohio. In May 1945 he assumed command of the Air Technical Service Command. He retired in the grade of major general. He was awarded the Distinguished Service Medal and Legion of Merit and was rated a senior pilot, combat observer and technical observer.

Air Force records end when Meyers was stripped of his $550 a month pension and decorations following conviction on 15 March 1948, in federal court on three counts of subornation of perjury. He was dismissed from the service of the United States by President Truman on 16 July 1948, after conviction. Once praised as the 'ablest procurement man in the Air Force whose efforts shortened the war,' dapper Benny was found to have secretly owned Aviation Electric Co. of Vandalia, Ohio. The perjury charges arose from his efforts to have the company's president, Bleriot LeMarr, lie to Senate investigators looking into siphoning of procurement funds into the firm. Meyers served just under three years of a twenty months-to-five year sentence when he was paroled from the federal reformatory at Lorton, VA. After some tax difficulties, he disappeared into private life.

The inevitable partisan dissension delayed release of the Senate committee's report of the Hughes investigation until April 1948, and even then it was only fully approved by the Republican members of the committee: Owen Brewster of Maine, Homer Ferguson of Michigan, Joseph McCarthy of Wisconsin, John Williams of Delaware, George Malone of Nevada, and Harry Cain of Washington.

The main conclusions of the report regarding the HK-1 flying boat were that the project, '...which produced no planes during the war, was an unwise and unjustifiable expense as a wartime project. The manpower, facilities, and public funds devoted to it during the war were wasted at a time when military planes were urgently needed. The conclusion is inescapable that the decision of the War Production Board was influenced because of the wide and favorable public acceptance of the proposal of Henry J. Kaiser for the mass production of huge cargo planes which Kaiser claimed would overcome the existing submarine menace to ocean transportation.' The report went on to state that '...the technical side of the war cannot be waged from day to day in a manner to accord with public opinion.'

The failure to follow normal procurement channels was, the committee concluded, '...a costly mistake. The Defense Plant Corporation did not have personnel qualified to supervise an aircraft construction program. Because of this inadequacy in personnel, the Civil Aeronautics Administration and the National Advisory Committee for Aeronautics, each were given some supervisory authority .... This divided authority, together with the inefficient management of the Hughes organization resulted in allowing Hughes Aircraft Company to carry on the project in an inefficient and wasteful manner'.

The following month, Democratic minority members of the committee, Carl Hatch of New Mexico, Claude Pepper of Florida, J. Howard McGrath of Rhode Island, and Herbert R. O'Conner of Maryland, issued a dissenting report. They declared that '...Howard Hughes and his companies were *entitled to a positive finding by the committee, especially so far as fraud, corruption, and willful wrong-doing are concerned. There is absolutely nothing in the evidence which discloses any fraud, corruption, or wrongdoing on the part of Howard Hughes or his associates.'*

Following the hearings, Hughes wrote an open letter to 'The Men and Women of Hughes Aircraft Company,' in which he reassured his employees that although he had just bought RKO he was not turning his back on the flying boat or on Hughes Aircraft. It seems he had high hopes for both: *'It is true that I have spent a good deal of time recently on the RKO deal.*

However, I am sure that you are also aware that I finished my design work in connection with the changes on the flying boat some time ago. There has intervened, by necessity, a period during which these changes are going through the process of final engineering, fabrication, assembly, and installation.

'I have tried to finish all phases of the RKO deal during this period, so that I will be free before the changes on the flying boat are completed.

'In view of the vicious political campaign which has been waged against me by certain competitive airline interests, together with their pal, Senator Brewster, and which attack has centered around the flying boat, it should hardly be necessary for me to say that nothing in this world means more to me than this airplane.

'And if we do finally succeed in designing, building, and flying an airplane twice as large as anything else in the world, and overcoming the hundreds of serious obstacles which are a part of this tremendous step ahead in the world's progress in aviation, then I believe people will wake up to what we have accomplished and complete vindication will have to follow.

'What's more important, we ourselves will know what we have done for aviation and I believe that one of these days this company will be one of the leading aircraft organizations of the United States, second to none in research, and recognized as such.

'When that time comes we can all hold our head up, and the more so because the way hasn't been easy and there have been plenty of people shooting at us.'

Fact Is Stranger Than Paranoia!

The question of was Senator Brewster really blackmailing Hughes, or did Hughes just think he was, is one that rolls on with the passing of time.

Senator Pepper, who was heavily involved in both the hearings and Washington politics of the time, later was to recall that *'...Brewster might as well have been on the payroll of Pan American as a vice president. He rode their planes as if he were an executive of the company. He called on them for all sorts of favors, and in addition to that he passionately fought their cause. I was on the committee of commerce where the decision had to be made as to whether we were going to support the chosen instrumentality legislation* [that would name Pan American the sole U.S. air carrier], *and Brewster had worked on me. He'd tried to get me to go along with it. I didn't have the slightest confidence in Brewster and I thought Hughes was telling the truth.*

Brewster was a strange kind of man. He had great energy, had a lot of ability, a lot of drive, but he never was regarded as a great man by his colleagues, or an outstanding politician because he was always involved in petty causes like this, pursuing them with a passion that suggested, whether it was confirmed or not, that he had a personal interest in the matter.'

Did Brewster's activities play on and inspire Hughes' passion - many would say paranoia - for secrecy regarding real or imagined threats on his privacy? By all accounts Hughes was a very different person before than after the Senate Hearings. A myriad of books all play on this to a great extent, as does the 2004 Martin Scorsese movie *'The Aviator'*.

An early account of this desire to take action to protect his secrecy came from Noah Dietrich, who had been amused by Hughes acting out the apparent fantasy of being hounded by 'them' by insisting on driving through the night streets of Washington discussing strategy at the time of the Senate Hearings because he feared electronic eavesdropping. Later, in 1949, it came to light that both Hughes' room at the Carlton and Dietrich's at the Mayflower had indeed been bugged!

Drew Pearson, who as a columnist was well used to the ways of Washington muck-raking and the manner of the DC two-step, had provided Hughes with much tactical advice in the Senate Hearings. In 1949, he let it slip that in a future echo to Watergate and the mysterious 'Deep Throat', he knew that police lieutenant Joe Shimon had done the job.

No one actually accused Brewster, but this was certainly Drew Pearson's assumption. Unlike later on with President Nixon and the whole Watergate affair, at the time not one of the appropriate Congressional committees was prepared to investigate a colleague.

Nevertheless, Pearson persuaded the sometimes receptive Senator Pepper to convene a sub-committee of the District of Columbia Committee on the premise that it was probing the police department, not Brewster.

This probe revealed a number of interesting facts:

1   Brewster had requested the Washington Police Department to assign to him Lieutenant Shimon, the department's wiretap expert

2   Shimon had received 'expense payments' from Brewster's secretary - a violation of police regulations.

3   In setting up the tapping and bugging operation, Shimon had instructed the three cops who manned the equipment to listen particularly for any references to Hughes, TWA or anything about airlines.

4   The eavesdropped information was transmitted to one Henry Grunewald.

5   Shimon was seen on one occasion receiving a $1,000 cash payment from a man fitting the description of Grunewald.

Senator Pepper is on record of saying that *'...Drew was a little unhappy that I wasn't more demanding and more exacting, but I told him that I went as far as I thought I could go on it. I didn't have any special ax to grind. I was simply trying to bring out the facts. And we brought out the facts. But what could I do with Brewster? I didn't want to bring a citation against him or anything. We'd shown him up before the public as engaged in that nefarious activity which the public had come to believe was par for the course for Brewster.'*

To Sleep.... To Dream.

In 1948 Hughes spent $1.75 million for a specially built, humidity controlled hangar where he kept the flying boat under twenty-four-hour guard. It was to sleep in this location for the next thirty-three years.

Over the years, the steady pumping of oil from under Terminal Island had caused the land to sink up to eighteen inches a year. The land surrounding the Hughes site had been built up by soil dredged from the channel; but the Hughes site had not been filled in, and only coffer dams and dykes kept the mud and water out. In 1953, the inevitable happened – the water broke through.

Hughes came down the night of the flood and looked everything over. The aircraft was floating and had been lifted against the overhead, damaging the props and tail. Scaffolding was hanging from all the nacelles. It was a mess - everyone was wading in the water, trying to get a diver down to get the debris out so the flying boat could be put back down in its cradle.

Hughes reportedly just said, '*Well, it's a hell of a mess. Let's get it cleaned up.*'

Over Noah Dietrich's urging to scrap it, Hughes repaired the flying boat - and then installed a series of expensive items to preclude damage from other such incidents ever happening again. Originally, the aircraft sat flat on the bottom of its dock on blocks, but as the island sank, in 1954 a team working around the clock in twelve hours shifts built up the sides of the dock and raised the machine on a cradle. Then an earthquake hit. There was no damage, but it shook everyone up.

The cradle was hooked up to a battery of compressed gas canisters that would be triggered if accidental flooding of the dock were to occur. These canisters were to inflate bags that would lift the nose and depress the tail so that the tail would not hit the overhead. The cradle could also be tilted six degrees nose up and six degrees nose down by huge walking beams moved by heavy cables that ran through pulleys to connect to the pistons of very large hydraulic cylinders on the floor of the dry dock.

Operational maintenance was performed continuously. Engines were run weekly until devices to rotate the engines and circulate hot pickle oil were designed and installed. The flight controls were exercised once a week. All the time every system was operated in accordance with an established schedule. Over the years many modifications were designed and installed. For years the crew fully expected that it would fly again.

Yet Hughes never flew the airplane again. Some, such as Noah Dietrich thought that Hughes was afraid of it. Others said that the flying boat had weaknesses. According to one of Hughes's mechanics, '*Maybe one of the reasons why they didn't fly it was there was a little fluctuation in the tail, and maybe it wasn't beefed up enough to suit him.*'

Another mechanic said, '*There was a lot of little damage. I don't know whether I should say it or not, but there was. That's when they went out into the wing and put little metal stiffeners in there to hold the glue joints*

together. A lot of angles snapped loose.'

Then there was an article in *The Chicago Tribune* for 5 May 1948 when reporter Wayne Thomis wrote:*'Through a spokesman, Hughes said here today that preparations are under way to sheath the hull and the wing-at least the half which carries the eight 3,500 horsepower engines of the 200-ton craft-with corrugated aluminum.*

'The metal is being applied to add stiffness and strength to the plywood structure. Need for this additional strength apparently showed up on the taxiing and 30-second flight last fall. We are devising entirely new methods for getting instant responses from our controls, both aerodynamic and power.'

'It should be understood clearly that the Hercules is considered only a research aircraft. It will never be used in competition with military or commercial machines, but will, in my estimation, be worthwhile because problems which will concern really big airplanes of the future will be solved here. The Hercules will point the way for really good big airplanes.'

Only Hughes knew why he never flew the boat again. Some thought that the reasons as to why Hughes never flew the plane again was physical problems brought on by the YF-11 crash and an escalation of his mental issues that just got worse and worse.

Rae Hopper said in one interview that he believed that other interests took up too much of his time. New tests were scheduled, but Hughes always cancelled out at the last minute. *'He just would not spend the time*

Photographs of the H-4 inside the hangar are incredibly rare, as everyone involved - an estimated 300+ employees were all sworn to secrecy.

with us that was needed to get ready for the next flight. I can't believe he was afraid. I wasn't.'

Others directly involved with the flying boat believed the same - that other interests and demands on Hughes's time kept him from flying the boat again. As one mechanic said *'When he first left the flying boat he was mixed up with RKO. After that came TWA. He just never got unmixed up with other stuff to come back to us.'*

A legend that did the rounds at one time was that Hughes had discussed with his close friend Cary Grant plans for a movie featuring the flying boat in which a secret agent (*a la* James Bond) would do battle for the good guys in exotic parts of the world. The flying boat would provide the agent's base and international mobility, just as 007 made use of advanced modes of transport.

It was probably little more than conversation. When Senator Ferguson asked Hughes during the hearings, *'Have you contemplated using this in your movie production business?'* Hughes replied, *'Senator, that is rather an absurd suggestion'*

One contributory factor that did make a lot of sense was that after World War Two the expected need for and interest in large flying boats simply dried up –thus giving Hughes no urgency to proceed with further tests.

The construction of military airfields in nearly every corner of the world during the war removed the incentive to develop flying boat docking and handling facilities, although Hughes did spend months designing a floating dock for the H-4.

Another factor was the nearly universal switch from wood to metal construction and the coming of the jet age. Hughes's wooden, propeller-driven flying boat was without doubt a tremendous achievement, but it was on the wrong side of a great watershed in aviation development.

In 1949 the General Services Administration took over the assets of the Reconstruction Finance Corporation including title to the flying boat. For the GSA the aircraft became a problem of disposal. Not only did the GSA want to get rid of the monster but the City of Long Beach wanted to regain its valuable waterfront property to develop a major oil terminal. In addition, the Internal Revenue Service no longer agreed to Hughes's use of flying boat expenses as a tax writeoff. Hughes stood to lose millions if he did not get rid of it.

Although the GSA had legal title to the flying boat and with that the right to sell it to anyone, the only people who wanted to buy it were promoters. Hughes said that if the boat were sold to 'any of their ilk,' he would sue. To him the flying boat was a great accomplishment and should be recognised as such, so the flying boat remained tucked away from prying eyes on Terminal Island.

Slowly, behind the scenes a deal was being struck. The GCA were talking to the Air Space Museum of the Smithsonian Institution in Washington DC, with the intention of giving the boat to them as a gift, but the Air and Space Museum had no use for it and no budget to move across the continent. It was also completely impractical to set up a Museum for one aircraft on the West Coast. What eventually emerged was a plan for the Smithsonian to accept the flying boat from the GSA, and then in turn the Smithsonian would give full title to it to the Summa Corporation (the umbrella corporation for Hughes enterprises) in exchange for the Hughes racer, which they desperately wanted, as the racer was regarded as a very important aircraft and the right size for their collection.

But in order for the government to come out at least even, it was decided that a group of aircraft appraisers and a committee of experts picked by the American Institute of Astronautics and Aeronautics would appraise both machines and decide what would be a fair difference. Eventually it was decided that the basic trade was that The Air and Space Museum got the Hughes racer and $700,000 from Summa, who then got clear title to the flying boat.

While all this was going on, a staff of 300 workers, reduced to 50 in 1962, kept the aircraft in flying condition.

There were plans that the flying boat would be donated to the Smithsonian and eight other museums around the country to be displayed in pieces. The Smithsonian would take the cockpit and a section of one wing and the remaining parts would be parceled out to the others, but vigorous public protest on the West Coast followed when the proposals to dismantle the huge flying boat were made public and the Summa Corporation was picketed.

On 15 April 1975, Carl Byoir & Associates issued a press release from Summa saying that the Smithsonian Institution and Summa had agreed to delay any dismantling of the aircraft. *'Although it is contemplated that a non-profit museum might devise some method of preserving and displaying the HK-1 intact, there are clearly many uncertainties as to costs involved in moving such a large craft overland, providing a suitable building, and operations and maintenance.'*

Towards the end of 1975 Summa contacted Rear Admiral Carl Seiberlich who was managing an Advanced Naval Vehicle Concept Evaluation Project for the Deputy Chief of Naval Operations for Air

The hangar for the flying boat on Terminal Island was a conglomeration of 'boxes' that were of a vague aircraft shape. Slowly the land on which it was built sank, and the surrounding ground was built up.

Warfare. The Summa representative told the admiral that they wanted to do something useful with the aircraft - something Mr. Hughes would have felt right about. Was the Navy interested?

Seiberlich agreed to look into it, and also to coordinate with other government agencies.

In November, a government team inspected the aircraft at Long Beach. They found a complete, airworthy machine that had a finish as smooth as glass. There was a complete set of engines and several sets of spares preserved in cans.

The Admiral explained why the Navy was interested. *'In this particular design, the hull and the solid struts to the tip floats act to fence in the main wing and engine areas and enhance the wing in ground effect. The propellers are all within this envelope. The wing flaps extend completely across it. Thus the flaps act to direct the prop blast and wing downwash trapped between these end plates downward towards the water surface. It would not take too much modification to make this an ideal wing and ground effect vehicle.'*

As part of their evaluation of advanced naval vehicle concepts, the Navy was interested in the phenomena - the extra lift provided when the wing's downwash cushions out against a ground or water surface. This effect extends to an altitude equal to about fifty per cent of the wingspan - one reason for the Navy interest in the 320-foot wing of the giant.

The Navy's idea was that if a plane were big enough, and designed to maximise the wing and ground effect, it would make possible transoceanic flights within the ground effect envelope at thirty per cent less power than would be required to fly comparable weights outside the ground effect range. Fuel savings would permit even greater loads to be carried.

Hughes's chief aerodynamicist, Carl Babberger: *'Classical ground effect occurs when the ground plane* [surface] *is near enough to the wing that it doesn't allow the air to be deflected downward as much as at altitude, and therefore the induced drag is less.'*

According to Dr. Harvey Chaplin, Director of the Aerodynamic Laboratory, David Taylor Model Basin, who is involved with wing and ground effect studies at the Pentagon, *'Carl Babberger is quite right. The kind of benefit you'd realize in flying transatlantic within the ground effect altitude would be a drag reduction which would reduce fuel consumption, increase range, and increase payload. This other business about the lift being generated from the propwash: we've been doing research on airplanes that generate a hell of a lot of lift that way. Generally, it depends on the gap through which you allow the air to escape underneath the flaps.'*

Thus there were very real reasons for Navy interest. As far as the public was concerned this is the only giant seaplane in existence; another like it would never be built. Admiral Seiberlich: *'So before we cut this one up, we want to be sure that we know all we need to know about large seaplanes.'*

What the admiral did not say was that US intelligence experts had spotted a huge, unknown craft on satellite reconnaissance photos of the Caspian Sea area. With its short wings, this vehicle dubbed 'Caspian Sea Monster' looked airplane-like in planform, but would obviously be incapable of flight. Although it was designed to travel a maximum of 3 metres above the sea, it was found to be most efficient at 20 metres,

The Soviet Ekranoplan KM, also known as Caspian Sea Monster. This was biggest and heaviest ekranoplan. It was 100 metres long and it weighed unbelievable 544 tons. The US Navy wanted to use to H-4 to discover more about the WIG concept.

reaching a top speed of 350 mph to 460 mph in research flights.

The Soviet 'ekranoplan' Wing In Ground effect (WIG) programme continued with the support of Minister of Defence Dmitriy Ustinov. It produced the most successful ekranoplan so far, the 125-ton A-90 Orlyonok. These craft were originally developed as high-speed military transports, and were usually based on the shores of the Caspian Sea and Black Sea. The Soviet Navy ordered 120 Orlyonok-class ekranoplans, but this figure was later reduced to fewer than 30 vessels, with planned deployment mainly in the Black Sea and Baltic Sea fleets.

This is probably the main reason why the U.S. Navy became interested in a full flight test programme for the 30-year-old Hughes flying boat. Unfortunately, after a six-month study, the Armed Services, NASA, and other interested agencies reluctantly concluded in March, 1977 that even though the aircraft was flyable, a test program with the flying boat would take money from projects of greater current priority.

Dreams Become A Nightmare

When Howard Hughes died 5 April 1976 without leaving a signed will, his estate was placed in the courts. In order to preserve the assets for Hughes' heirs, the courts ruled that all unnecessary expenses be curtailed. In Long Beach, the H-4 was still being preserved at a cost of more than $1 million a year. Summa Corporation, Hughes' holding company, was ordered to take steps to dispose of the aircraft and so let the remaining fifty workers employed on the site go.

Summa took over the billionaire's assets, but could not find a new home for the aircraft, so they reluctantly planned to disassemble the aircraft and dispose of the pieces. Nine museums, including the Smithsonian, would receive parts of the plane for display. The remainder of the craft would be shredded and destroyed. Across the nation, thousands of aviation enthusiasts were horrified. In 1979, a California-based group, The Committee to Save the Flying Boat, staged demonstrations and rallies and called for support to save the aircraft from destruction. Letters and phone calls deluged local and state politicians, demanding the craft's preservation.

The Port of Long Beach briefly examined the possibility of displaying the Hughes Flying Boat, but again, money was a problem: millions of dollars

Arelo Sederberg, vice president of Carl Byoir & Associates, speaking for Summa Corporation in April 1977 said that preliminary discussions had been held with a group of aviation history buffs who had formed the Air Museum of the West. Long Beach Municipal Judge Gilbert Alsten, a museum organiser, was optimistic about raising the several million dollars required. *'We're hoping we can save the airplane. Nothing like it has ever been built before, and nothing like it probably will ever be built again.'* Their organisation sparked a great deal of interest in the project, but in the end they lacked the necessary backing to consummate a deal.

In January 1980, the Summa Board of Directors met at corporate headquarters in Las Vegas and postponed a decision as to whether to put the Hughes Hercules on display next to the *Queen Mary* in Long Beach or at the Aero World theme park planned for Mira Mesa near San Diego, or at a theme park near Redwood City, California on San Francisco Bay called Marineland-Africa-USA.

Aero World developers would require that the aircraft be disassembled and barged to San Diego. The Redwood City site was such that the airplane would not have to be disassembled, but could be moved intact by sea up the coast, under the Golden Gate and San Francisco-Oakland Bay Bridges to the bayside display site.

The move to a site near the *Queen Mary* in Long Beach would be easiest of all. This, plus the attractions of a double bill, gave Long Beach an advantage if only the proper mix of money, organisation, and management could be found.

But there were problems. Long Beach had lost money on the *Queen Mary* project and were reluctant to become over-committed. Bill Berry, former Summa program manager for the flying boat, went on record saying that Long Beach did not want to put up any money. *'They wanted us to pay for the advertising, for exhibits, carpeting, and so on. We felt that wasn't in our original bargain, that that is part of museum operation.*

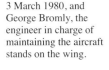

3 March 1980, and George Bromly, the engineer in charge of maintaining the aircraft stands on the wing.

Another view inside the hangar - a view from the roof, that somehow shows just how deep the hull was in the dock.

Our offer was to provide a display building and pay for the delivery of the airplane.'

In early April 1980, the Committee to Save the Flying Boat under Chairman Robert L. McCaffery held a press conference aboard the *Queen Mary* after Summa had announced that if $175,000 could be put up to move the flying boat, Summa would fund the rest. But by the time the Summa board met to consider the problem on 21 May, the money had not been raised. Summa announced plans to dismantle the boat.

A 23 May 1980, *Los Angeles Times* article reported that *'...the world's largest airplane, Howard Hughes' 'Spruce Goose' plywood flying boat, will be cut up and distributed to nine museums...'*

McCaffery's committee continued its media blitz and contacted local Assemblyman Bob Dornan, an ex-fighter pilot. Dornan sparkplugged a resolution of the State of California declaring the aircraft to be of historic significance.

The efforts of the Committee to Save the Flying Boat attracted national attention which culminated in the appearance of Glenn Odekirk and Robert McCaffery on the 16 July 'Good Morning America' TV show and with the listing of the flying boat in the National Register of Historic Places of the Department of the Interior.

In the meantime, Richard Stevens, an active member of the Aero Club of Southern California and president of Wrather Hotels, helped bring the Wrather Corporation and the Aero Club into the picture. Wrather Hotels operated Disneyland Hotel, among other of its concerns, as part of Jack Wrather's Beverly Hills-based entertainment conglomerate.

In July 1980, the Aero Club developed a letter of intent, jointly signed by Summa, Wrather, and Nissen Davis, president of the Aero Club of Southern California and vice president for public relations of Flying Tiger Airline. The letter of intent stated that upon the satisfaction of certain conditions, Summa would donate the flying boat to the Aero Club, a nonprofit organisation, which would enter into a management agreement with Wrather to move the machine, design and erect a permanent building to display the boat, and operate a museum. The funds required for the move and the building would total more than $2.5 million and will be provided by loans to the Aero Club from Summa of $1.5 million, Wrather of $250,000 and loans from ARCO, and others. The money would be repaid from expected receipts from the operation of the museum and associated activities.

At last it appeared that a viable combination of money, organisation, personnel, connections, and experience had been found. On 25 August 1980, Wrather signed a forty-year lease to operate the RMS *Queen Mary* and display the Hughes flying boat in Long Beach Harbor and began an aggressive programme of development and promotion.

The First Move

The H-4 would be housed in an appropriate building on land adjacent to the RMS *Queen Mary*, but an immediate problem surfaced: the flying boat would have to be moved at once because its hangar had already been leased to another company. In September 1980, the job of moving the aircraft began. Major settling - nearly twelve feet - had occurred on Terminal Island over the years. This meant that when in water the flying boat would be much higher than when the hangar was originally built. In addition, the rolling doors of the hangar were jammed shut and would not open.

Work crews used cutting torches to remove the doors and all parts of the hangar that might touch the aircraft when it was floated out – that took nearly two months.

February 1982, and the H-4 is barged across Long Beach Harbor into the new geodetic dome alongside RMS *Queen Mary.*

The flying boat was finally removed from its hangar on 29 October 1980 after the building that was erected around it was partially dismantled around it.

The move called for the aircraft to be towed to a storage site just 600 yards north of the hangar facility, where the aircraft would be lifted on to land by the world's largest self-propelled floating crane, the Navy's YD-171 - known in Long Beach as *Herman the German,* as it was a war prize taken from Germany at the end of World War Two and was the only crane on the Pacific Coast that could have lifted the 400,000-pound craft.

The cradle on which the flying boat had rested for decades would be removed from the dry dock, attached to lifting beams and placed underwater. The flying boat would then be towed into place over the cradle, adjacent to the YD-171 and cables would be lowered into the water and attached to the sunken cradle by divers. The crane would then lift the cradle up underneath the plane's fuselage, where divers would ensure proper alignment before the airplane was lifted into the air and carried across the narrow channel to land.

The move did not go smoothly. Two mobile cranes, trying to lift the cradle out of dry dock, began to sink into the floor of the hangar. One tipped over, losing its boom in the dry dock. The Terminal Island subsidence had so badly undercut the floor of the hangar that the cranes could not get sufficient footing. It took many hours to shore up the floor so that the crane could finally lift the cradle out of the dry dock. When the cradle was free, it was placed on a special barge where enormous lifting beams were welded to the underside of the cradle. The steel cables from the YD-171 were attached to these lifting beams.

The original schedule called for the move to take six hours, but sunset fell before the cradle was prepared.

For two days the H-4 floated in Long Beach Harbor before it was lifted onto a temporary storage area on shore.There it remained for something over a year for its new display location on Pier J to be prepared. The same day, on 31 October William Rice Lummis, representing the Hughes family and Summa, officially turned the flying boat over to the Aero Club.

The Wrather Corporation had selected Temcor of Torrance, California, to build the aluminum dome to cover the H-4. At 415 feet in diameter and 130 feet in height it covered more than three acres and was the largest clear-span aluminum dome in the world.

The H-4 on display in the Dome alongside the RMS *Queen Mary.* *(Photo: David Lee)*

Two more views of H-4 on display in the dome alongside the RMS *Queen Mary.*

A mannequin of Hughes sits at the control just as he did on the one and only flight.

John Morgan shows his strength by holding up the tail of the H-4. *(Photos: David Lee)*

Over the night of 10/11 February 1982 it was lifted from its storage location onto a barge. The next morning it was floated across the harbour to a giant geodesic dome to be displayed by the Wrather Corporation, illuminated by a multi-colored light show. It was displayed alongside a replica of the HR-1 racing plane, and a Howard Hughes look-alike wandered around the dome, greeting visitors.

When Michael Eisner became chairman and CEO of Walt Disney Productions in 1984, he wanted to get out of Disney's agreement with the Wrather Corporation and bring the Disneyland Hotel under the Walt Disney corporate umbrella. Wrather continued to refuse to sell to Eisner, just as he had refused Walt Disney many years before. Wrather died two months after Eisner took over at Disney, and four years later, in 1988, Disney bought the entire Wrather company. At the time Wrather's company also owned the RMS *Queen Mary* and the lease on the H-4 Hercules as well as the rights to *The Lone Ranger* and the *Lassie* TV series. Though Disney kept the hotel, it has disposed of a large number of other assets that came with the purchase.

The Second Move
In 1992, Evergreen Aviation & Space Museum co-founders Michael King Smith and Delford M. Smith submitted the winning proposal to provide the aviation icon with a proper home and subsequently bought the aircraft from the Aero Club.

Disassembly of the airplane began on 10 August for its transportation to another new home in Oregon. The Flying Boat was disassembled and transported by barge up the West Coast, then down the Columbia and Willamette Rivers, to Portland, Oregon - a journey by ship, barge and truck that took 138 days and covered 1,055 miles. It remained there for several months, until water levels permitted the huge structures to safely pass under the Willamette's many bridges.

In February 1993, the aircraft was transported by truck for the last

Just as they turned out to watch the flying boat move from Culver City to Long Beach in 1947, thousands turn out to see the arrival of the H-4 in McMinville in 1993.

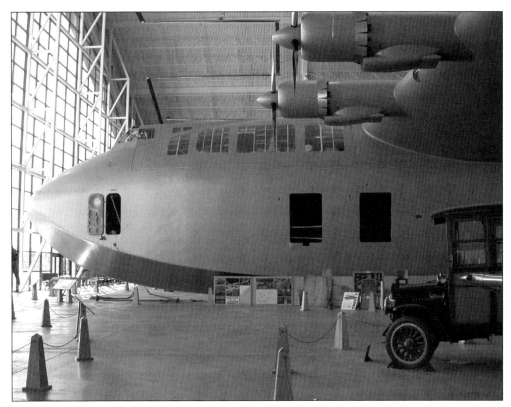

The H-4 on display at McMinville. The extra holes cut into the hull date back to the days of being on display at Long Beach - they were cut to allow visitors inside.

The upper deck glazed over holes were to allow the visitors to see inside the flight deck area from a walkway.

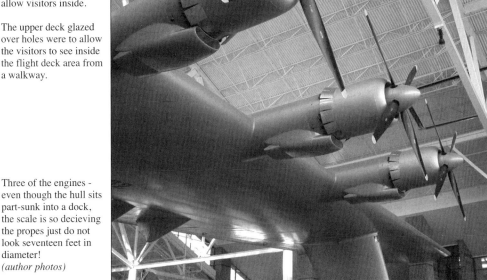

Three of the engines - even though the hull sits part-sunk into a dock, the scale is so decieving the propes just do not look seventeen feet in diameter!
(author photos)

seven and a half miles to McMinnville, Oregon. Temporary hangars were built as housing for the aircraft, while volunteers worked on the aircraft's restoration.

Construction on the museum was planned to begin almost immediately after the flying boat arrived in McMinnville but was soon put on hold.

Evergreen International, the main contributor to the effort, fell on temporary hard times forcing a delay early on in the project. Work was ready to begin again when tragedy struck. Delford Smith's son, Michael Smith was killed in a tragic accident. He was serving as the president of the museum at the time of his death. The project was again halted while the family dealt with it's loss and the organization regrouped. The museum was then dedicated to the memory of Capt. Michael King Smith and the group was more determined than ever to complete the museum.

For several years the Hughes Flying Boat remained disassembled in a shelter awaiting the opening of what was now called the The Captain Michael King Smith Evergreen Aviation Educational Institute.

The components of the flying boat were transported to the new museum site in September of 2000. It has been reassembled and is now on display. It has received a new coat of silver paint, so it looks like it did when it made its only flight.

To get on board
So what is it like to actually get on board Hughes' pride and joy?

My first thought on going through the hatch was 'this thing is big!' – not A380 or 747 big, because on boarding those airliners they are full of seats, interior trim, overhead lockers all fitted in compartments – The Hercules is much more 'open'.

The next thing that strikes you is the craftsmanship apparent in the construction of this piece of living history. The joinery of the laminated wood is so precise, it appears to be one piece instead of the many layers that it is. The overwhelming impression is of being inside a massive wooden musical instrument as I looked down the length of the fuselage – it's like you are standing inside a huge Stradavarius violin!

Though larger than if made of aluminum, the ribs still look more like metal than wood. This aircraft takes structural engineering to a level that is close to being art, and the workmanship simply has to be seen to be appreciated! From the main deck a narrow spiral staircase – that was originally a vertical ladders - leads to the flight deck. Once there, it's past the engineering stations where the many systems of the aircraft were monitored. The instruments are still in place on the tables where they sat during the flight. Approaching the place where Howard Hughes controlled this beast all those years ago I was motioned towards the left seat; '...*go ahead, sit down, see how it feels*'.

Getting into the seat was no problem, after all this is a very roomy flight deck! The view out the front is surprising, the nose is not visible at all from boss's chair, sloping away at an angle so steep you can't see it. For some reason the control wheel seems small for such a large machine, but with the massive hydraulic powered controls, no more was needed. Eight tall power levers dominate the area between the pilot's well-appointed station and the co-pilot's sparsely instrumented panel.

I could imagine the roar of the eight Pratt & Whitney R-4360 radial

'My first thought on going through the hatch was 'this thing is big!' *(author photos)*

engines would have made that day.... daydreaming for a moment, I took in all that was around me, sitting in this special place of aviation history, I was certain that the excitement of that moment those years ago still manages to live on here....

In late 2013 it was revealed that the McMinville Museum was facing financial scrutiny, and the ownership of the historic airplane was in question. The Evergreen Aviation and Space Museum backed by Evergreen International Airlines founder Delford Smith, 83, and along with some of Smith's other business entities, the cargo carrier was facing financial difficulty. According to reports in a number of Oregon news sources, creditors were trying to force Evergreen International Airlines into bankruptcy, and the Oregon Department of Justice was investigating whether or not funds from Smith's businesses have been illegally co-mingled with museum financing.

Though the popular belief is that the museum bought the H-4 for $1, but apparently Smith agreed to pay its previous owner $500,000 over 20 years, plus a share of the museum profits. An attorney for the Aero Club of Southern California claimed the museum still owed $50,000. Museum director Larry Wood told reporters he was unaware of the debt, and Smith himself said in a telephone interview, *'I had no idea that they felt we still owed them money. I thought we were free and clear on* [the H-4].'

While it is possible the Aero Club could repossess the flying boat, its fate would be in question. It took a massive and expensive effort to dismantle and move the aircraft. The climate-controlled building housing the airplane was essentially constructed around it. Aero Club attorney Robert Lyon said, *'To call it a white elephant would be an understatement. But concern has arisen of what we are going to do if they default on the note and we have to take the boat back. I wonder if they have enough money to take it apart and get it out of the building.'*

And so the saga of the H-4 Hercules continues.

Technical Description

The Hughes Flying Boat is a cargo type airplane having eight Pratt and Whitney R4360-4A engines, single vertical tail, fixed wing tip floats, and full cantilever wing and tail surfaces. Its structure and surface are entirely of laminated wood; all primary control surfaces except the flaps are fabric covered.

Contained in the hull are a flight deck for the operating crew and a large cargo deck with a circular stairway providing access from one deck to the other. Below the cargo deck are located the fuel bays separated by watertight bulkheads. Beaching of the airplane is accomplished by the use of dry dock facilities.

Dimensions

(a) Wing and Empennage -

	Wing	Horizontal Tail	Vertical Tail
Area in square feet:	11430.0	2610.1	1703.443
Span in feet:	319.92	113.5	49.5
Mean Aerodynamic Chord in feet:	38.13	24.681	38.231
Root Chord in feet:	51.7	31.13	53.167
Tip Chord in feet:	19.65	16.808	17.144
Root Section:	NACA 63, (420) 321, (a = 1)	NACA 0012-64 (modified) to 10. 714% thickness	
Tip Section:	NACA 65, (318) 415, (a = 1)	NACA 0012-64(modified) to 10. 714% thickness	
Geometric Twist:	4°	0°	0°
Aspect Ratio:	9.0	0.5399	0.322
Incidence with fuselage reference line:	+5° (at root cord)	+1 0°	0°
Dihedral:	3-1/2°	0°	0°

(b) Control Surfaces

	Flaps	Elevators x2	Rudder	Ailerons x2
Total area in square feet:	1404.0	1110.200	650.560	1250.0
Area aft of hinge line in square feet:	-	832.800	477.924	780.0
Span in feet:	-	104.318	47.661	143.5
Span in % of wingspan:	46.0	-	-	-
Average chord aft of hinge line in feet:	-	7.98	10.0	5.45
Chord (% of surface chord):	22.84	34.2	28.3	20.0
Control travel:	45°	15° down 25° up	20° left 20° right	12°down 18° up
Type of nose balance:	-	overhang	overhang	internal seal

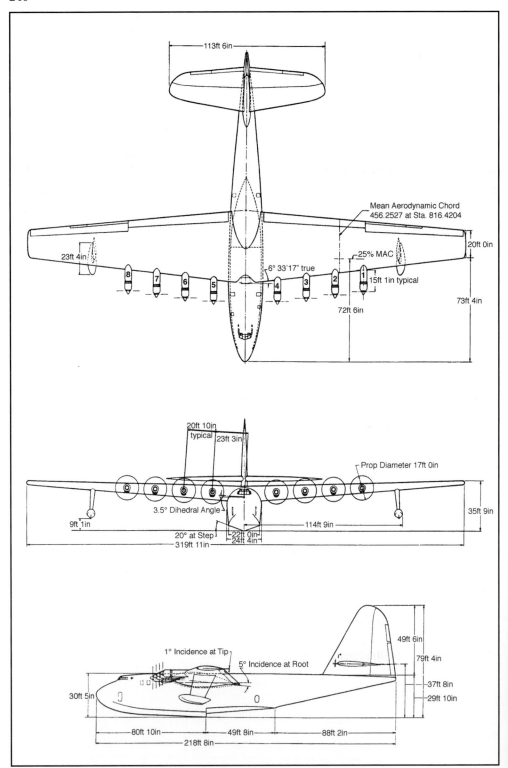

(c) Flying Tabs	Elevator	Rudder	Aileron
Area after of hinge line in square feet:	90.28	57.834	97.0
Span in feet:	46.80	23.069	62.08
Average chord after of hinge line in feet:	1.90	2.50	1.58
Chord (% of control surface chord):	20	20	25
Tab travel:	9° down	15° left	15° down
	15° up	15° right	15° up

(d) Trim Tabs	Elevator	Rudder	Aileron
Area aft of hinge line in square feet:	48.28	22.503	33.6
Span in feet:	15.526	11.875	25.0
Mean Geometric Chord:	1.58	1.9	1.35
Chord (% of control surface chord):	20	20	25
Tab travel:	15° up	15° up	15° up
	15° down	15° down	15° down

Performance

	Gross Wright 350,000 lb	Gross Weight 400,000 lb
High speed at sea level with Take off power in mph:	235.5	234
High speed at sea level with normal rated power in mph:	222	218
High speed at 5000ft with normal rate power in mph:	231	227
Operating speed at sea level with max cruising power (1675 BHP/2230 rpm) in mph:	190	185
Initial cruising speed for max range at 5000ft altitude:	141	150
Landing speed at sea level - 10% above stall, 45° flap:	81	87
Range in miles at best cruising speed with no fuel reserve		
12, 500 gallons of fuel:	-	2975
6,000 gallons of fuel:	1575	1430
Endurance in hours at best cruising speed		
12, 500 gallons of fuel:	-	20.9
6,000 gallons of fuel:	11.5	9.5
Service Ceiling in feet:	20,900	17,400
Rate of climb at sea level with normal rated power in feet/minute:	878	675
Rate of climb at sea level with Take Off power in feet/minute:	1,134	916
Time to climb to 5000ft altitude with normal rated power in minutes:	5.9	7.7

Bibliography

A Rich Young Texan with a Poet's Face Gets Hero's Welcome on World Flight. Life magazine. July 25, 1938.

Aircraft Contracts, Hughes Aircraft Co. (Washington, D.C.: Government Printing Office, 1948).

Aircraft Engineering Division, Civil Aeronautics Authority, letter to Special Aviation Assistant to the Secretary of Commerce, January 20, 1944, *War Production*, File 314.444, RG 179, US National Archives.

Aircraft Production; the journal of the aircraft manufacturing industry monthly, 1944

Aircraft Production; the journal of the aircraft manufacturing industry monthly, 1945

Aircraft Production; the journal of the aircraft manufacturing industry monthly, 1946

Aircraft Production; the journal of the aircraft manufacturing industry monthly, 1947

Aircraft Production; the journal of the aircraft manufacturing industry monthly, 1948

Assistant Deputy Commanding General, Research & Development, Procurement & Industrial Mobilization to Chief of Staff United States Air Force, Director of Public Information, Legislative and Liaison Division, October 8, 1947. Record Group 18, "Records of the Army Air Force," *Files of Assistant Chief of Air Staff, Materiel & Services, Research & Development Branch, Case Histories 1941-1946*, D-2, D-5, F-11 Project, Vol. 11, US National Archives.

Aviation Week Howard Hughes article August 11, 1947.

Aviation Week Howard Hughes article, August 11, 1947.

Aviation Week, Howard Hughes article April 26, 1948.

Chief of Aircraft Engineering, Flight Engineering and Factory Inspection Divisions, letter to Director, Safety Regulations, Civil Aeronautics Authority, September 7, 1943. *War Production* File 314.4442, RG 179, US National Archives.

Citizen Hughes Michael Drosnin Hutchinson & Co. London. 1985

Col. R.B. Lord, CE., USA, to Chairman, War Production Board, August 31, 1942, *War Production Board* File 314.444, RG 179, US National Archives.

Confessions of a Muckraker Jack Anderson with James Boyd, (New York: Random House, 1979),

Design Aspects of the Hughes H-4. Industrial Aviation, October 1945.

Donald M. Nelson letter to Howard R. Hughes. February 8, 1943, 314.444, RG 179, NA.

Drew Pearson, Diaries: 1949-1959, ed. by Tyler Abell (New York: Holt Rinehart & Winston, 1974),

E.A. Locke, Jr. to Donald M. Nelson, August 31, 1942, *War Production* File 314.444, RG 179, US National Archives.

Edward R. Murrow, *I Can Hear It Now,* Columbia

Masterworks Recording, XLP 1774, Side 1.

Empire: The Life, Legend and Madness of Howard Hughes. - Donald L. Barlett and James B. Steele. W.W. Norton & Company, New York 1979. ISBN 0-393-07513-3

Facing Death With Howard Hughes John Leyden. Horizons (Washington, D.C.: Federal Aviation Administration)

George W. Lewis to Grover Loening, January 27, 1944, Record Group 255, *Records of the National Advisory Committee for Aeronautics,* US National Archives.

Grover Loening memo to Chairman, War Production Board, February 22, 1943, *War Production,* File 314.444, RG 179, US National Archives.

Grover Loening memo to Donald Nelson, August 21, 1943, *War Production* File 314.4442, RG 179, US National Archives.

Grover Loening memo to George W. Lewis, January 27, 1944, *War Production* File 314.444, RG 179, US National Archives.

Grover Loening memo to Nelson, *Notes on meeting February 29th with Howard Hughes*. March 2, 1944, File 314.4446, RG 179, US National Archives.

Grover Loening memo to Nelson, September 10, 1943, *War Production* File 314.4442, RG 179, US National Archives.

Grover Loening memo to Wilson, *Report on visit to Hughes Aircraft Company and recommendations on HK-1 project* September 29, 1943, *War Production* File 314.4442, RC 179, US National Archives.

Grover Loening report of August 2, 1943 visit to Kaiser-Hughes Aircraft plant, August 21, 1943, Record Group 179, *Records of the War Production Board* File 314.4442, National Archives Building, Washington, D.C.

Grover Loening to Donald M. Nelson, August 28, 1942, Record Group 179, *Records of the War Production Board,* File 314.4442, National Archives Building, Washington, D.C.

Grover Loening to Nelson, summary report of meeting held February 18, 1944, dated February 19, 1944, *War Production* File 314.44463, RG 179, US National Archives.

Grover Loening to Wilson, November 29, 1943, *War Production* File 314.444, RG 179, US National Archives.

Grover Loening, *Comments on Preliminary General Specifications for Hughes-Kaiser Cargo Airplane Model HK-1* April 3, 1943, Record Group 179, *Records of the War Production Board* File 314.44463, RG 179, National Archives, Washington, D.C.

H.H. Arnold, memorandum for the Under Secretary of War, July 17, 1944.

Harold E. Talbott letter to Donald M. Nelson, March 4, 1944, File 314.4442, RG 179, NA.

Hearings. 80th Congress, lst session. Part 40, Aircraft

Contracts (Hughes Aircraft Co. and Kaiser-Hughes Corp.) (Washington, D.C.: Governent Printing Office, 1947). Part 43, Aircraft Contracts, Hughes Aircraft Co. (Washington, D.C.; Government Printing Office, 1948).

Houston Post Howard Hughes article July 31, 1938

Howard Hughes John Keats. New York: Random House, 1972

Howard Hughes and his Flying Boat - Charles Barton Aero Publishers, Fallbrook, CA: 1982. Republished in 1998, Vienna, VA: Charles Barton, Inc. ISBN 0-9663175-0-5.

Howard Hughes and pseudoaddiction. Practical Pain Management

Howard Hughes and the Spruce Goose. John J McDonald. Tab Books Inc., Blue Ridge Summit, Pennsylvania: 1981. ISBN 0-8306-2320-5.

Howard Hughes Breaks His Silence. History.com

Howard Hughes Documentary. Amazon.

Howard Hughes letter to Donald Nelson, March 18, 1943, *War Production* File 314.4442, RG 179, NA.

Howard Hughes Revealed. National Geographic Channel, Inside (series), Season 7, episode 2.

Howard Hughes, Airman Extraordinaire. FAA World, January 1977.

Howard Hughes, Don Dwiggins Plane & Pilot, May 1971

Howard Hughes: A Chronology. Channel 4 TV

Howard Hughes: Aviator. George J. Marrett Naval Institute Press, Annapolis, Maryland: 2004. ISBN 1-59114-510-4.

Howard Hughes: The Real Aviator. Amazon.

Howard Hughes: The Secret Life, Charles Higham, 1993.

Howard Hughes: The Untold Story. - Peter Harry Brown and Pat H. Broeske. Penguin Books, New York 1996. ISBN 0-525-93785-4.

Howard Hughes. John Simkin. Spartacus Educational.

Howard Hughes. U.S. Centennial of Flight Commission, 2003.

Howard Hughes's Amazing Aircraft Part 3 and 4. Aeroplane Monthly July/August 1974

Howard R. Hughes, Jr. - The Record Setter. David H Onkst. U.S. Centennial of Flight Commission, 2003.

Howard: The Amazing Mr. Hughes. Noah Dietrich and Bob Thomas. Fawcett Publications, New York 1972. ISBN 00449025651.

Hugh Fulton to E.A. Locke, Ir., January 14, 1944, *Aircraft, Cargo-Production*. Record Group *179, Records of the War Production Board* File 314.444, National Archives, Washington, D.C.

Hughes Alters Plane to Strengthen It. Chicago Tribune, May 4, 1948.

Hughes Answers Senate Probe With First Flying Boat Flight. Aviation Week, November 10, 1947.

Hughes Flying Boat, World's Largest Airplane. Automotive and Aviation Industries, August 15, 1945

Hughes H-4 'Spruce Goose'. Concept Aircraft: Prototypes, X-Planes and Experimental Aircraft. Jim Winchester. Grange Books plc., Kent, UK: 2005. ISBN 978-1-59223-480-6.

Hughes letter to Nelson, March 17, 1944, File 314.4446, RG 179, US National Archives.

Hughes: The Private Diaries, Memos and Letters: The Definitive Biography of the First American Billionaire. Richard Hack.: New Millennium Press, Beverly Hills, California 2002. ISBN 1-893224-64-3.

Jesse Jones to Donald Nelson, February 16, 1944, *War Production* File 314.4446, RG 179, US National Archives.

John W. Snyder, Executive Vice President Defense Plant Corporation to Henry]. Kaiser and Howard Hughes, September 17, 1942, *War Production* File 314.4442, RG 179, US National Archives.

Kaiser telegram to Donald Nelson, February 16, 1944, *War Production* File 314.4446, RG 179 NA.

Loening memo to C.E. Wilson, Executive Vice Chairman, War Production Board, September 29, 1943, *War Production* File 314.4442, RG 179, NA.

Lt. Col. H.A. Freidlick, Chief, Contracts and Facilities Division, Memorandum for the File, Iuly 2, 1944, Files of Assistant Chief of Air Staff Materiel and Services (A-4) Research and Development Branch, Case Histories 1941-1946, Airplane F-11, Record Group 18, *Records of the Army Air Forces* National Archives, Washington, D.C.

Mammoth Hughes H-4 Nears Takeoff Line. Aviation, October, 1945

Memorandum for the files on meeting in Donald Nelson's office, October 21, 1943, dated October 25, 1943, *War Production* file 314.4441 RC 179 National Archives.

Memorandum, FH. Hoge, Jr. to Chairman, Planning Committee, War Production Board, May 22, 1942. File 314.444, Aircraft, Cargo- Production, Record Group 179, Records of the War Production Board, National Archives, Washington, D.C.

Minutes of Kaiser-Hughes Flying Boat Conference, NACA Conference Room, May 4, 1943, *War Production* File 314.444, RG 179.

Minutes, Aircraft Production Board meeting, October 4, 1943, *War Production* File 314.44-42, RG 179, NA.

Minutes, Aircraft Production Board, November 29, 1943, File 314.4442, RG 179, NA.

Minutes, Aircraft Production Board, October 25, 1943, *War Production* File 314.444, RG 179, NA.

Minutes, Kaiser-Hughes Flying Boat Conference, NACA Conference Room, May 14, 1943, *War Production* File 314.444, RG 179, NA.

New York Times Howard Hughes article August 4, 1947

New York Times Howard Hughes article August 6, 1947.

New York Times Howard Hughes article July 30, 1942.

New York Times Howard Hughes article August 30, 1979.

New York Times Howard Hughes article,]uly 29, 1947

New York Times Howard Hughes article, July 8, 1946

New York Times Howard Hughes article, February 1, 1947

New York Times Howard Hughes article, July 11, 1942

New York Times Howard Hughes article, November 16, 1936

New York Times Howard Hughes article August 10 1947

Newsweek Howard Hughes article August 4, 1947.

Newsweek Howard Hughes article, August 18, 1947.

Next to Hughes: Behind the Power and Tragic Downfall of Howard Hughes by his Closest Advisor. Robert Maheu and Richard Hack. Harper Collins, New York: 1992. ISBN 0-06-016505-7.

Obituary: Robert Maheu: FBI agent and CIA fixer who became Howard Hughes's bagman. The Guardian, August 20, 2008

Paramount News, August 13, 1947, PN 6.100, Record Group 200, National Archives, Washington, D.C.

Paramount News, November 8, 1947, PN 7.21, Record Group 200, National Archives, Washington, D.C.

R. Adm. J.H. Towers to Donald Nelson, September 12, 1942, and Robert A. Lovett to Donald Nelson, September 12, 1942, *War Production* File 314.4442, RG 179, US National Archives.

R.L. Horne memo to files regarding October 4, 1943 meeting in Nelson's office, *War Production* File 314.4442, RG 179, US National Archives.

Seaplanes & Flying Boats: A Timeless Collection from Aviation's Golden Age. Bill Yenne. New York: BCL Press, 2003. ISBN 1-932302-03-4.

Speed Merchant Airpower, September 1977, Walt Boyne

Spruce Goose (Title inside cover: HK-1 Hercules: A Pictorial History of the Fantastic Hughes Flying Boat) Glenn E Odekirk. Glenn E. Odekirk and Frank Alcantr, Inc., Long Beach, California: 1982.

Spruce Goose Evergreen Aviation Museum.

Supervising Engineer, Defense Plant Corporation, Culver City, California. Memo Re: Kaiser-Hughes-Plancor 1424, Management of Cargo Plane Project, August 27, 1943, *War Production* File 314.4442, RG 179, US National Archives.

Telecon log, Chad Calhoun in Washington, Henry Kaiser in New York, and Donald Nelson, January 29, 1944, *War Production* File 314.4446, RG 179, US National Archives.

Telecon log, Howard Hughes and Donald Nelson, February 17, 1944 *War Production* File 314.44463, RG 179, US National Archives.

Telecon log, Hughes and Nelson, February 18, 1944, *War Production* File 314.44463, RG 179, US National Archives.

The American Aviator: The Howard Hughes Story. Vision Films.

The American Heritage History of Flight (New York: American Heritage Publishing Company, Inc., 1962)

The Asylum of Howard Hughes. Jack Real. Xlibris Corporation, Philadelphia: 2003. ISBN 1-4134-0875-3.

The Beauty and the Billionaire. Terry Moore. Pocket Books, New York: 1984. ISBN 0-671-50080-5.

The Hughes H-4, Biggest Wooden Airplane Ready. Canadian Aviation, October 1945

The Oregonian Howard Hughes article. Portland, Ore., July 20, 1942.

The Passions of Howard Hughes. Terry Moore and Jerry Rivers. General Publishing Group, Los Angeles: 1996. ISBN 1-881649-88-1.

The Reminiscences of Roscoe Turner, May 1960, Aviation Project, Volume III, Part 3, Oral History Collection, Columbia University.

The Spruce Goose Commemorative Pictorial. Milton L Schwartz. The Wrather Corporation Oakland, California: by Mike Roberts Color Productions, 1983.

Time Magazine Howard Hughes article, August 18, 1947

Time Magazine Howard Hughes article, August 4, 1947.

Times Herald Howard Hughes article, Newport News, Virginia April 27, 1977.

Tycoons: The Secret Life of Howard Hughes Time, magazine December 13, 1976

Untitled, unsigned memo, February 3, 1944, *War Production* File 314.4446, RG 179, US National Archives.

W. L. Clayton to Donald Nelson, February 17, 1944, *War Production* File 314.4446, RG 179, US National Archives.

War Production File 314.444, Record Group 179, Robert E. Gross to Brig. Gen. Charles E. Branshaw, August 12, 1942, with copy to Donald M. Nelson.

Washington Post Howard Hughes article, August 12, 1947.

Washington Post Howard Hughes article, August 7, 1947..

Washington Post Howard Hughes article, August 8, 1947.

Washington Post Howard Hughes article, August 8, 1947.

Wilson to Loening, December 8, 1943, File 314.444, RG 179, US National Archives.

Index